International Association of Fire Chiefs

Fire Officer
Principles and Practice
SECOND EDITION

Student Workbook

JONES AND BARTLETT PUBLISHERS

Sudbury, Massachusetts

BOSTON   TORONTO   LONDON   SINGAPORE

Jones and Bartlett Publishers
World Headquarters
40 Tall Pine Drive
Sudbury, MA 01776
978-443-5000
info@jbpub.com
www.jbpub.com

Jones and Bartlett Publishers Canada
6339 Ormindale Way
Mississauga, Ontario
L5V 1J2
Canada

Jones and Bartlett Publishers
  International
Barb House, Barb Mews
London W6 7PA
United Kingdom

National Fire Protection Association
1 Batterymarch Park
Quincy, MA 02169
www.NFPA.org

International Association
  of Fire Chiefs
4025 Fair Ridge Drive
Fairfax, VA 22033
www.IAFC.org

Jones and Bartlett's books and products are available through most bookstores and online booksellers. To contact Jones and Bartlett Publishers directly, call 800-832-0034, fax 978-443-8000, or visit our website, www.jbpub.com.

Substantial discounts on bulk quantities of Jones and Bartlett's publications are available to corporations, professional associations, and other qualified organizations. For details and specific discount information, contact the special sales department at Jones and Bartlett via the above contact information or send an email to specialsales@jbpub.com.

Production Credits

Publisher: Kimberly Brophy
Senior Acquisitions Editor—Fire: William Larkin
Associate Managing Editor: Robyn Schafer
Associate Production Editor: Sarah Bayle
V.P., Manufacturing and Inventory Control: Therese Connell
Composition: diacriTech
Cover Design: Kristin E. Parker

Cover Image: © Mark C. Ide
Printing and Binding: Courier
Cover Printing: Courier

Editorial Credits

Author: Douglas C. Ott

ISBN: 0-7637-8367-9

6048
Printed in the United States of America
13 12 11   10 9 8 7 6 5 4 3 2

Contents

*Note to the reader: Exercises indicated with a 🔥 are specific to the Fire Officer II level.

Answer Key

Student Resources

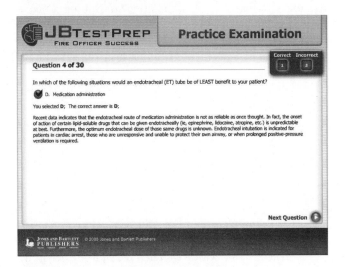

JBTest Prep: Fire Officer Success

ISBN-13: 978-0-7637-8369-3

JBTest Prep: Fire Officer Success is a dynamic program designed to prepare students to sit for Fire Officer I and/or II certification examinations by including the same type of questions they will likely see on the actual examination.

It provides a series of self-study modules, organized by chapter and level, offering practice examinations and simulated certification examinations using multiple-choice questions. All questions are page-referenced to *Fire Officer: Principles and Practice, Second Edition* for remediation to help students hone their knowledge of the subject matter.

Students can begin the task of studying for Fire Officer I and/or II certification examinations by concentrating on those subject areas where they need the most help. Upon completion, students will feel confident and prepared to complete the final step in the certification process—passing the examination.

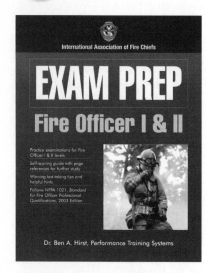

Exam Prep: Fire Officer I & II
ISBN-13: 978-0-7637-2761-1

Exam Prep: Fire Officer I & II is designed to prepare you to sit for a Fire Officer I and II certification, promotion, or training examination by including the same type of multiple-choice questions you are likely to encounter.

To help improve examination scores, this preparation guide follows Performance Training Systems, Inc.'s Systematic Approach to Examination Preparation®. *Exam Prep: Fire Officer I & II* is written by fire personnel explicitly for fire personnel, and all content has been verified with the latest reference materials and by a technical review committee.

Benefits of the Systematic Approach to Examination Preparation include:

- Emphasizing areas of weakness
- Providing immediate feedback
- Learning material through context and association

Exam Prep: Fire Officer I & II includes:

- Practice examinations for Fire Officer I and Fire Officer II levels
- Self-scoring guide with page references for further study
- Winning test-taking tips and helpful hints
- Coverage of NFPA 1021, *Standard for Fire Officer Professional Qualifications*

Technology Resources

www.Fire.jbpub.com

This site has been specifically designed to complement *Fire Officer: Principles and Practice, Second Edition* and is regularly updated. Resources available include:

- **Chapter Pretests** that prepare students for training. Each chapter has a pretest and provides instant results, feedback on incorrect answers, and page references for further study.

- **Interactivities** that allow students to reinforce their understanding of the most important concepts in each chapter. Interactivities include:
  - **You Are the Fire Officer**
  - **Fire Officer In Action**
  - **Skill Objectives Activities**

- **Hot Term Explorer**, a virtual dictionary, allowing students to review key terms and test their knowledge of key terms through quizzes and flashcards.

Introduction to the Fire Officer

Workbook Activities

The following activities have been designed to help you. Your instructor may require you to complete some or all of these activities as a regular part of your fire officer training program. You are encouraged to complete any activity that your instructor does not assign as a way to enhance your learning in the classroom.

Chapter Review

The following exercises provide an opportunity to refresh your knowledge of this chapter.

Matching

Match each of the terms in the left column to the appropriate definition in the right column.

_____ 1. Fire mark   **A.** The guidelines that a department sets for fire fighters to work within.

_____ 2. Rules and regulations   **B.** An officer responsible for a single fire company or shift.

_____ 3. Fire chief   **C.** Formal statements that provide guidelines for present and future actions.

_____ 4. Lieutenant   **D.** An amplification device once used to give orders during an emergency.

_____ 5. Discipline   **E.** Developed by government-authorized organizations to implement a law that has been passed by a government body.

_____ 6. Chief's trumpet   **F.** The process of identifying problems and opportunities and resolving them.

_____ 7. Policies   **G.** A symbol used to identify homes protected with fire insurance in the mid 1700s.

_____ 8. Decision making   **H.** The individual assigned the responsibility for management and control of all matters and concerns of the fire department.

_____ 9. Deputy chief   **I.** A process used to accomplish an objective.

_____ 10. Planning   **J.** The officer in charge of a major functional area, for example, training.

Multiple Choice

Read each item carefully and then select the best response.

_____ 1. The first step in a progressive sequence generally associated with an officer supervising a single fire company is the rank of:
   A. captain.
   B. lieutenant.
   C. Fire Officer I.
   D. Fire Officer II.

_____ 2. The professional qualifications standards for fire officers are documented in NFPA:
  A. 1001.
  B. 1021.
  C. 1035.
  D. 1041.

_____ 3. The duties of both Fire Officer I and Fire Officer II can be divided into administrative, emergency, and:
  A. disciplinary activities.
  B. public relations activities.
  C. operations activities.
  D. nonemergency activities.

_____ 4. A department that uses full-time career personnel along with volunteer or paid-on-call personnel is referred to as a:
  A. matched department.
  B. combination department.
  C. rural department.
  D. mixed department.

_____ 5. The formal rank structure of most fire departments was adopted from the:
  A. Roman military.
  B. British military.
  C. Greek military.
  D. American military.

_____ 6. In the United States, a fire department somewhere is responding to a fire every:
  A. 10 seconds.
  B. 20 seconds.
  C. 30 seconds.
  D. 60 seconds.

_____ 7. In response to the Peshtigo Fire and Great Chicago Fire, communities began to enact:
  A. fire brigades.
  B. fire laws.
  C. strict building and fire codes.
  D. insurance policies.

_____ 8. For fire officers to effectively coordinate firefighting efforts, it is vital that they have good:
  A. communication.
  B. response times.
  C. equipment.
  D. experiences.

_____ 9. The most important resources on the fire scene are the:
  A. water storage system.
  B. transportation vehicles.
  C. rescue and suppression equipment.
  D. fire fighters.

_____ **10.** Building codes govern the types of construction materials and, frequently, the built-in fire prevention and:
    A. safety measures.
    B. insurable limits.
    C. occupancy loads.
    D. fire resistance levels.

_____ **11.** The fire chief is directly accountable to the:
    A. local government.
    B. county or district governing body.
    C. state or provincial government.
    D. federal government.

_____ **12.** Captains report directly to:
    A. lieutenants.
    B. chief officers.
    C. assistant chiefs.
    D. fire chiefs.

_____ **13.** The chain of command ensures that given tasks are:
    A. completed in a timely manner.
    B. appropriately delegated.
    C. carried out in a uniform manner.
    D. documented by department members.

_____ **14.** The complex process of influencing others to accomplish a task is the act of:
    A. management.
    B. planning.
    C. discipline.
    D. leadership.

_____ **15.** Written organizational directives that establish specific operational methods to be followed routinely for the performance of designated operations are:
    A. rules and regulations.
    B. policies.
    C. standard operating procedures.
    D. standing orders.

_____ **16.** Developing clear organizational values can be accomplished by:
    A. having a code of ethics that is well known throughout the organization.
    B. selecting employees who share the values of the organization.
    C. ensuring that top management exhibit ethical behavior.
    D. All of the above.

Fill-in

Read each item carefully, and then complete the statement by filling in the missing word(s).

1. The emergency duties of the Fire Officer II include supervising a multi-unit emergency operation using a(n) _____.

2. The two most common forms of staffing in fire department organizations are _____ and _____.

3. The _____ is a part of management and is responsible for the conduct of others.

4. The first organized volunteer fire company was established under the leadership of _____.

5. In Canada, every _____ minutes a fire department responds to a fire.

6. _____, a fire fighter in New York City, developed the first fire hydrants in 1817.

7. _____ are charged with protecting the welfare of the public against common threats.

8. Ethical choices are based on a(n) _____ system.

Short Answer

Complete this section with short written answers using the space provided.

1. At your first training session as a fire officer, you notice that several of your former crew members are not conducting themselves appropriately. How do you handle the situation?

_____

_____

_____

_____

2. You are the first fire officer on scene at a large structure fire. What steps do you take to establish command?

_____

_____

_____

_____

3. As a fire officer, how will you approach your duties differently from when you were a fire fighter?

_____

_____

_____

_____

4. Which personal challenges do you need to be aware of during the transition from fire fighter to fire officer?

_____

_____

_____

_____

5. Why do you want to be a fire officer?

_____

_____

_____

_____

6. What is your personal code of ethics?

_____

_____

_____

_____

Fire Alarms

The following case studies will give you an opportunity to explore the concerns associated with becoming a fire officer. Read each case study and then answer each question in detail.

1. Today is your first day as a new lieutenant. You are both excited and apprehensive at the same time, unsure of what the day will bring. When you enter the station, you receive several phone calls and requests from a few of the fire fighters now assigned to your shift. This is something new and you suddenly realize that you're in a different position within the chain of command. What is the importance of that chain of command and where do you fit within it as a lieutenant?

_____

_____

_____

_____

_____

_____

2. Fire Fighter Link has been seriously thinking about taking the upcoming lieutenant exam that was recently posted. In the past, the department has held promotional exams, but the department has not promoted anyone from the exams in over five years. Fire Fighter Link wonders if it is even worth it to take the test. However, after speaking with Lieutenant Sobol, he is told that there are at least seven officers who will be retiring soon. Why is this occurring?

_____

_____

_____

_____

_____

In-Basket

The following exercise will give you an opportunity to refine your writing skills.

1. Develop presentation notes for new members outlining the fire department organizational chart. Include a brief description of position duties and specific individuals from the fire fighter to local governing body levels.

   _____

   _____

   _____

   _____

   _____

   _____

2. Prepare a brief description of the duties assigned to your fire department (Fire Officer I and Fire Officer II level officers).

   _____

   _____

   _____

   _____

   _____

   _____

Preparing for Promotion

Workbook Activities

The following activities have been designed to help you. Your instructor may require you to complete some or all of these activities as a regular part of your fire officer training program. You are encouraged to complete any activity that your instructor does not assign as a way to enhance your learning in the classroom.

Chapter Review

The following exercises provide an opportunity to refresh your knowledge of this chapter.

Matching

Match each of the terms in the left column to the appropriate definition in the right column.

_____ 1. Personal study journal

A. An attribute or quality that can be described and measured during a promotional examination.

_____ 2. Job description

B. A promotional question that provides an opportunity for the candidate to demonstrate depth of knowledge in a particular subject.

_____ 3. Dimensions

C. A worksheet used to provide a map and identify the factors needed to be evaluated in a promotional examination.

_____ 4. Class specification

D. A narrative summary of the scope of a job.

_____ 5. Assessment centers

E. A technique used to assess persuasiveness.

_____ 6. Emergency incident simulations

F. The traits required for every classified position within the municipality.

_____ 7. "Data dump"

G. A technique used to assess problem analysis and decision-making skills.

_____ 8. KSAs

H. Used to test a candidate's ability to perform the role of an officer at a fire or other type of situation.

_____ 9. Oral presentation

I. A notebook to aid in scheduling and tracking a candidate's promotional preparation progress.

_____ 10. Written report

J. A series of simulation exercises used to evaluate a candidate's competence in performing the actual tasks associated with the job.

Multiple Choice

Read each item carefully and then select the best response.

_____ 1. In response to the abuse of government job promotions, the civil service system in the federal government was established by the:
A. Patronage Act.
B. Pendleton Act.
C. Congressional Act.
D. Fire Services Act.

_____ 2. A metropolitan fire department is an agency with more than:
   A. 200 fully paid fire fighters.
   B. 300 fully paid fire fighters.
   C. 400 fully paid fire fighters.
   D. 500 fully paid fire fighters.

_____ 3. The emphasis of the promotional examination is determined by the classified job description and the:
   A. rank.
   B. community need.
   C. department size.
   D. timeline for hiring.

_____ 4. When promotional candidates are grouped into categories of "Highly Qualified," "Qualified," and "Not Qualified," the department utilizes a(n):
   A. grouping list.
   B. cluster approach.
   C. evaluation scoring system.
   D. banded list.

_____ 5. Most promotional job announcements require successful candidates to have an appropriate:
   A. educational background.
   B. physical performance rating.
   C. number of years of service.
   D. driving abstract.

_____ 6. The preparation of a promotional examination usually involves the combined effort between the fire department and the municipality's:
   A. financial section.
   B. maintenance section.
   C. human resources section.
   D. town administrators.

_____ 7. Examples of the typical tasks that a person holding that position would be expected to perform is provided in the:
   A. job specification.
   B. class specification.
   C. KSA description.
   D. job description.

_____ 8. The first-level supervisory examination usually includes a focus on:
   A. management duties.
   B. administrative procedures.
   C. technical questions covering rescue company operations.
   D. budget proposals.

_____ 9. "Book smart, street dumb" was a phrase often used in complaints about promotional processes that relied solely on:
   A. oral presentations.
   B. written multiple-choice exams.
   C. written reports.
   D. performance evaluations.

_____ **10.** A typical in-basket exercise contains instructions, exercise items, and:
   A. a calendar.
   B. a policy manual.
   C. purchase orders.
   D. a service manual.

_____ **11.** Promotional candidates have to react to unfolding situations in:
   A. emergency incident simulations.
   B. oral presentations.
   C. training scenarios.
   D. interpersonal interaction exercises.

_____ **12.** When candidates are placed into groups, this is called a:
   A. ranking list.
   B. banded list.
   C. breakdown list.
   D. candidacy list.

_____ **13.** In general, the first-level supervisory examination has the:
   A. longest examination.
   B. greatest number of topics.
   C. most diverse reading list.
   D. shortest examination.

_____ **14.** When developing examinations, smaller departments are more likely to:
   A. go to an external source.
   B. develop an examination specific to their department.
   C. hire a consultant.
   D. increase the interview question list and avoid the written examination.

_____ **15.** An interpersonal interaction exercise is designed to test a candidate's ability to:
   A. communicate with a group.
   B. review and prioritize a list of tasks.
   C. deal with disciplinary issues.
   D. perform effectively as a supervisor.

_____ **16.** The assessment center process was developed in the 1920s by:
   A. an American fire chief.
   B. the officer corps of the German army.
   C. the FBI.
   D. the Federal Employee Testing Association.

Fill-in

Read each item carefully, and then complete the statement by filling in the missing word(s).

1. Prior to the Civil War, most government jobs were awarded based on personal relationships or the
_____ system.

2. Jurisdictions have the ability to make promotions to meet departmental and community _____.

3. The promotional process is constantly _____ to meet community needs and political challenges.

4. Human resources departments prepare a technical _____ worksheet to quantify the KSA components
of municipal jobs.

5. The _____ examinations are widely used and can be structured to focus on specific subjects and factual
information.

6. A(n) _____ exercise is a common assessment center event that involves the candidate dealing with a
stack of correspondence that has accumulated in a fire officer's in-basket.

Short Answer

Complete this section with short written answers using the space provided.

1. What is the promotional examination process used in your department?

_____

_____

_____

_____

2. What opportunity for advancement exists in your department? When is the next opportunity?

_____

_____

_____

_____

3. What are the narrative job description and technical class specifications for fire officers in your department?

_____

_____

_____

_____

4. Identify your two strengths and greatest challenges when considering your department's promotional process.

_____

_____

_____

_____

5. Identify the people who will help you prepare for the promotional process.

_____

_____

_____

_____

6. In reviewing your personnel file and history within the fire service, what are your strengths, weaknesses, and greatest learning moments?

_____

_____

_____

_____

Fire Alarms

The following case studies will give you an opportunity to explore the concerns associated with becoming a fire officer. Read each case study and then answer each question in detail.

1. You have been a fire fighter for six years and are ready to take the next step in your career to become a lieutenant. An exam notice has been posted and you meet the minimum requirements to participate in the testing process. The announcement states the exam consists of a written exam, assessment center, and oral board. The assessment center will include an emergency incident simulation, in-basket exercise, and interpersonal interaction exercise. You have never taken an assessment center and do not know what any of the exercises consist of. How do you prepare for the assessment center?

_____

_____

_____

_____

_____

2. You work for a medium-sized metropolitan fire department with a workforce of 90 personnel. You have been eligible to take the fire officer exam for two years, but there have been no testing opportunities. You recently asked your battalion chief when the next testing opportunity was going to occur and he stated that the department had no plans to be testing soon. What influences a fire department to test for fire officer?

_____

_____

_____

_____

_____

In-Basket

The following exercise will give you an opportunity to refine your writing skills.

1. Prepare a response to a high school student's e-mail requesting information on the chain of command and career progression for fire fighters within your department.

_____

_____

_____

_____

_____

2. You have been selected to represent your department at the local high school's Career Day. The Chief has asked you to be prepared to explain the fire department's promotional process. Briefly describe the minimum qualifications required for each rank in your department.

_____

_____

_____

_____

_____

_____

Fire Fighters and the Fire Officer

Workbook Activities

The following activities have been designed to help you. Your instructor may require you to complete some or all of these activities as a regular part of your fire officer training program. You are encouraged to complete any activity that your instructor does not assign as a way to enhance your learning in the classroom.

Chapter Review

The following exercises provide an opportunity to refresh your knowledge of this chapter.

Matching

Match each of the terms in the left column to the appropriate definition in the right column.

_____ 1. Integrity

_____ 2. Consent decree

_____ 3. Walk around

_____ 4. Actionable items

_____ 5. Personal training library

_____ 6. Hostile work environment

_____ 7. Confidential

_____ 8. Diversity

_____ 9. Ethical behavior

_____ 10. Sexual harassment

A. A fire workforce that reflects the community it serves.

B. A three-ring binder used to keep notes and documentation from training events.

C. When an employee is subject to discrimination in the workplace.

D. Keeping information private.

E. Unwanted, uninvited, and unwelcome attention and intimacy in a nonreciprocal relationship.

F. A complex system of inherent attributes that determine a person's moral and ethical actions.

G. Employee behavior that requires immediate corrective action.

H. A legal remedy where the fire department agrees to accomplish specific goals within a specific time.

I. An informal, unscheduled tour of the fire station.

J. Behaviors that are consistent and reflect the department's core values, mission, and value statements.

Multiple Choice

Read each item carefully and then select the best response.

_____ 1. The form that is faxed from every fire station to the battalion chief within 15 minutes of the reporting time is the:
   A. shift report.
   B. daily status report.
   C. beginning of a shift.
   D. task report.

_____ **2.** Injury reports and other information that must be passed up the chain of command quickly are sent by:
  A. report.
  B. memos.
  C. personal contact.
  D. notification.

_____ **3.** To assist in scheduling and planning, the duty crew activities many departments use is a:
  A. day book.
  B. service log.
  C. status report.
  D. chore or duty list.

_____ **4.** Major changes in how the fire fighter relates to the formal fire department organization occur:
  A. twice during their career.
  B. four times during their career.
  C. annually.
  D. after each emergency incident.

_____ **5.** The direct responsibility for the supervision, performance, and safety of a crew of fire fighters belongs to the:
  A. command officer.
  B. lieutenant.
  C. battalion chief.
  D. company level officer.

_____ **6.** Typically the direct responsibility of several fire companies belongs to the:
  A. command officer.
  B. lieutenant.
  C. battalion chief.
  D. company officer.

_____ **7.** Supervisor, commander, and trainer are three distinct roles of the:
  A. safety officer.
  B. command officer.
  C. fire officer.
  D. battalion chief.

_____ **8.** The fire officer will supervise the fire company in a manner that is consistent with the rules and regulations of the:
  A. community's administration.
  B. fire department.
  C. NFPA standards.
  D. fire chief.

_____ **9.** Fire officers should strive to establish a working relationship with every building manager in their:
  A. response district.
  B. community.
  C. emergency plan.
  D. high risk report.

_____ **10.** If a fire officer's decision is going to have an impact that goes beyond the fire officer's scope of authority, he or she must:
   A. have support from administration.
   B. discuss with his or her platoon.
   C. report within the morning report.
   D. contact his or her supervisor.

_____ **11.** A steadfast adherence to a moral code and internal value system is a sign of:
   A. ethical behavior.
   B. honesty.
   C. integrity.
   D. accountability.

_____ **12.** Equal employment opportunity and affirmative action laws and department initiatives are in place to address:
   A. ethical behavior.
   B. employer equity.
   C. hostile workplaces.
   D. workplace diversity.

_____ **13.** The standard for evaluating sexual harassment is based on similar circumstances and reactions by a(n):
   A. reasonable person.
   B. adult.
   C. representative from the human resources department.
   D. jury.

_____ **14.** The fire officer's designated role in conducting an investigation of an EEO complaint depends on the:
   A. administrative policy.
   B. human resources department guidelines.
   C. fire department's procedures.
   D. severity of the complaint.

_____ **15.** The fire department has an agency-wide mission that is translated into:
   A. scheduled duty tasks.
   B. fire fighter conduct.
   C. annual goals.
   D. department guidelines.

_____ **16.** The term/phrase _____ is used to describe a broad range of situations in which an employee is subject to discrimination in the workplace.
   A. hostile work environment.
   B. firefighter harassment.
   C. hazing.
   D. tradition.

Fill-in

Read each item carefully, and then complete the statement by filling in the missing word(s).

1. The _____ report projects any staff adjustments at the beginning of a shift and the anticipated staffing for the next day.

2. The fire officer has to anticipate that _____ will alter the workday and will require adjustments to the schedule.

3. Fire officers must be able to and are expected to make _____.

4. After completion of initial training and the probationary period, many departments acknowledge the achievement with a change in helmet _____.

5. A fire fighter's relationship with the formal fire department changes primarily due to his or her sphere of _____ within the organization.

6. The formal fire department organization considers a(n) _____ to be the fire chief's representative at the work location.

7. A command officer has _____ of a hands-on role than a fire officer of a company.

8. If a fire officer has concerns or objections to an order, he or she should express them with his or her supervisor in _____.

9. Command _____ is the ability of an officer to project an image of being in control of the situation.

10. If a fire officer has the _____ to solve a problem, he or she should address and resolve the situation.

Short Answer

Complete this section with short written answers using the space provided.

1. What do you remember about being a probationary fire fighter?

_____

_____

_____

_____

2. How could your relationship with the other members of the fire department change when you become a fire officer?

_____

_____

_____

_____

3. As a fire officer, you overhear one of your fire fighters making derogatory and racist comments. What action and steps do you take?

_____

_____

_____

_____

4. What are your department's policies and procedures concerning filing an EEO/AA complaint?

_____

_____

_____

_____

5. How do you behave toward a fire fighter in your department who has filed an EEO/AA complaint?

_____

_____

_____

_____

6. How do you handle yourself in informal situations around the fire department when derogatory, racial, or off-color humor is used?

_____

_____

_____

_____

Fire Alarms

The following case studies will give you an opportunity to explore the concerns associated with becoming a fire officer. Read each case study and then answer each question in detail.

1. While on your shift, you are approached by one of your fire fighters who requests to speak with you privately as soon as possible. You oblige the request and invite her to meet with you in your office. She tells you that one of the other fire fighters has been making rude and inappropriate comments related to females in her presence. She asked the individual to refrain from the action in hopes that it would end. Unfortunately it only escalated. She now considers the conduct to be harassing and requests that you deal with the issue. If necessary, she will file a formal complaint. What should you do?

_____

_____

_____

_____

_____

2. After returning from a call, you hear several fire fighters talking about a memo that was recently posted by the assistant chief. You go and read the memo that has been posted on the memo board and find that it pertains to going to the grocery store during the shift. The fire fighters are angry about the memo and it upsets you as well. There was no discussion of the subject with any of the fire officers and you do not personally see a problem. You question the memo being placed on the memo board for public view and the reluctance of the assistant chief to bring it to the fire officers for discussion. Some members of the department are even threatening to take the issue to the union. Later that same day, you see the aid unit crew leaving the station. You ask them where they are going and they tell you that they are going to pick some items up at the store. You know that the fire fighters are ignoring the new memo. How should you respond to this situation?

_____

_____

_____

_____

_____

In-Basket

The following exercise will give you an opportunity to refine your writing skills.

1. Develop notes for a five-minute presentation to a general community meeting aimed at recruiting members for your volunteer fire department.

_____

_____

_____

_____

_____

_____

2. Prepare an answer to the following potential promotional interview question. You have been promoted to company officer and are assigned to the same station and company where you worked as a fire fighter. How will you handle the lack of respect your former coworkers may demonstrate toward you when you give assignments in the station?

_____

_____

_____

_____

_____

_____

Understanding People: Management Concepts

Workbook Activities

The following activities have been designed to help you. Your instructor may require you to complete some or all of these activities as a regular part of your fire officer training program. You are encouraged to complete any activity that your instructor does not assign as a way to enhance your learning in the classroom.

Chapter Review

The following exercises provide an opportunity to refresh your knowledge of this chapter.

Matching

Match each of the terms in the left column to the appropriate definition in the right column.

_____ **1.** Theory Y	**A.** Resulting from managers who believe that people do not like to work.
_____ **2.** Blake and Mouton	**B.** Supporting the idea that people like to work and that they need to be encouraged, not controlled.
_____ **3.** Human resources management	**C.** Theory X and Theory Y.
_____ **4.** Employee relations	**D.** The process of setting and evaluating an employee's performance against a standard.
_____ **5.** Theory X	**E.** The process of attracting, selecting, and maintaining an adequate supply of labor.
_____ **6.** Performance management	**F.** Hierarchy of needs.
_____ **7.** Staffing	**G.** Compensation system based on pay-per-grade level composed of a number of steps.
_____ **8.** Step-and-grade pay system	**H.** Managerial Grid Theory.
_____ **9.** Maslow	**I.** All activities designed to maintain a rapport with the employees.
_____ **10.** McGregor	**J.** Includes the functions of staffing, performance management, benefits, employee health, and safety.

Multiple Choice

Read each item carefully and then select the best response.

_____ **1.** The nature of work was changed in the late 1800s by optimizing the ways that tasks were performed so workers could be trained to perform a specialized function. This is the concept of:
   A. scientific management.
   B. human resources.
   C. motivation.
   D. employee effectiveness.

_____ **2.** Human resources management is built from theories based in:
  A. administration and operational management.
  B. scientific and humanistic management.
  C. efficiency and production management.
  D. Taylor and McGregor management.

_____ **3.** Due to the nature of emergency services, the second step of Maslow's hierarchy of needs is a primary concern because it addresses:
  A. safety and security.
  B. social affiliations.
  C. self-esteem.
  D. self-actualization.

_____ **4.** Promotions, awards, and membership on elite teams addresses Maslow's level of:
  A. safety and security.
  B. social affiliations.
  C. self-esteem.
  D. self-actualization.

_____ **5.** When a person feels whole, alive, and more aware of the truth, Maslow describes them as:
  A. balanced.
  B. efficient.
  C. self-actualizing.
  D. sound.

_____ **6.** Blake and Mouton's style that shows the lowest level concern for both results and people is the:
  A. status quo style.
  B. accommodating style.
  C. controlling style.
  D. indifferent style.

_____ **7.** Blake and Mouton's style that demonstrates a high concern for results, along with a low concern for others, is the:
  A. status quo style.
  B. accommodating style.
  C. controlling style.
  D. indifferent style.

_____ **8.** Blake and Mouton's style with the objective of playing it safe and working toward acceptable solutions is the:
  A. status quo style.
  B. accommodating style.
  C. controlling style.
  D. indifferent style.

_____ **9.** Blake and Mouton identify the preferred model for a candidate to become a successful officer as the:
- **A.** self-actualizing model.
- **B.** Theory Y model.
- **C.** humanist model.
- **D.** sound behavioral model.

_____ **10.** Human resources management focuses on the:
- **A.** support of administrative policy.
- **B.** task of managing people.
- **C.** development of operational efficiency.
- **D.** fire fighter's career planning.

_____ **11.** The more direct supervision that is needed, the:
- **A.** less efficient the crew.
- **B.** more efficient the crew.
- **C.** less operational the crew.
- **D.** more operational the crew.

_____ **12.** A file that uses a page to describe each activity and when it is to be completed is often referred to as a(n):
- **A.** task log.
- **B.** activity sheet.
- **C.** duty roster.
- **D.** daily file.

_____ **13.** Management as we know it today is a product of the:
- **A.** Industrial Revolution.
- **B.** Management Evolution.
- **C.** human resources department.
- **D.** legal system.

_____ **14.** The four principles of scientific management are credited to:
- **A.** Henry Ford.
- **B.** Douglas McGregor.
- **C.** Frederick Winslow Taylor.
- **D.** the Los Angeles City Fire Department.

_____ **15.** Information on laws that affect fire fighters can be found through the:
- **A.** Department of Emergency Services.
- **B.** Department of Labor.
- **C.** IAFF.
- **D.** NFPA.

_____ **16.** The psychologist best known for his research on mental health and human potential is:
- **A.** Douglas McGregor.
- **B.** Frederick Winslow Taylor.
- **C.** Abraham H. Maslow.
- **D.** George Elton Mayo.

Fill-in

Read each item carefully, and then complete the statement by filling in the missing word(s).

1. _____ is the science of using available resources to achieve desired results.

2. Direct supervision requires that the fire officer _____ observe the actions of the crew.

3. The _____ is a formal document that outlines the basic reason for the organization and how it sees itself.

4. The fire officer can ensure maximum _____ by using time management skills.

5. One of the best tools to improve efficiency is _____ because it allows subordinates to complete tasks they are capable of performing.

6. McGregor found that workers with greater _____ were more likely to be motivated in their jobs.

7. Blake and Mouton described _____ behavioral models based on a person's position on the managerial grid.

8. The _____-based pay systems pay a base salary and then give additional compensation for any skills the fire fighter gains or can demonstrate.

9. Retirement plans, paid holidays, sick leave, and life insurance are typical examples of the _____ of a compensation system.

10. The most basic need on Maslow's ladder is _____ because without it there is no motivation or energy to do anything but survive.

Short Answer

Complete this section with short written answers using the space provided.

1. As a fire fighter, can you recall a time when you wanted more freedom in completing your duties?

_____
_____
_____
_____

2. During your transition to fire officer, what steps will you take to build trust with your fire fighters?

_____
_____
_____
_____

3. As a fire officer, what tasks will you be able to delegate?

_____
_____
_____
_____

4. What can you do to promote employee responsibility and autonomy in your day-to-day operations and during emergency responses?

_____
_____
_____
_____

5. As a fire officer, what tasks will be a priority that were not when you were a fire fighter?

_____

_____

_____

_____

6. How can you assist in a fire fighter's career growth utilizing Maslow's hierarchy of needs?

_____

_____

_____

_____

7. What steps can you take to make your department operate more efficiently?

_____

_____

_____

_____

Fire Alarms

The following case studies will give you an opportunity to explore the concerns associated with becoming a fire officer. Read each case study and then answer each question in detail.

1. It is August and it is a very hot, dry day. Your crew has already made several runs this morning and you decide that the crew needs to catch up on some shift training that has been neglected this month. Part of the training involves completing three hose evolutions. After the second, you become very frustrated with the performance of your crew. They are very sluggish and their attitudes are less than desirable. You advise them that they will be out there until the evolutions are satisfactorily completed. No improvements are made. What should you do?

_____

_____

_____

_____

_____

2. You have been a fire officer for 10 years and during that time you have learned a large amount about yourself as a leader and those who you lead. You have learned the need for trust and respect and how they play a key role in your ability to be an effective fire officer. Recently, you observe a new fire officer assigned to your shift having a difficult time getting his crew to listen. You have observed this a couple times and have heard through the grapevine that there are trust issues with his crew. This new officer worked for you at one time and you believe that he will make an excellent officer with the right direction. What should you do?

_____

_____

_____

_____

_____

_____

In-Basket

The following exercises will give you an opportunity to refine your writing skills.

1. Develop a personal mission statement about your management style and philosophy.

_____

_____

_____

_____

_____

_____

2. Explain how understanding Theory X and Theory Y management concepts can help you become a better supervisor.

_____

_____

_____

_____

_____

_____

3. Make a scorecard of your positives and challenges with each style of management presented in this chapter.

_____

_____

_____

_____

_____

_____

Organized Labor and the Fire Officer

Workbook Activities

The following activities have been designed to help you. Your instructor may require you to complete some or all of these activities as a regular part of your fire officer training program. You are encouraged to complete any activity that your instructor does not assign as a way to enhance your learning in the classroom.

Chapter Review

The following exercises provide an opportunity to refresh your knowledge of this chapter.

Matching

Match each of the terms in the left column to the appropriate definition in the right column.

_____ **1.** Grievance

_____ **2.** Right-to-work

_____ **3.** Collective bargaining

_____ **4.** Political action committee

_____ **5.** Strike

_____ **6.** Grievance procedure

_____ **7.** Fair Labor Standards Act

_____ **8.** Binding arbitration

_____ **9.** Yellow dog contracts

_____ **10.** Impasse

A. When a group of employees withhold their labor for the purposes of effecting a change in wages.

B. The resolution of a dispute by a third and neutral party.

C. A special interest group that can solicit funding and lobby elected officials for their cause.

D. A worker cannot be compelled, as a condition of employment, to join or not to join, or to pay dues to a labor union.

E. Establishes minimum standards for wages and spells out administrative procedures covering work time and compensation.

F. A complaint that any employee may have in relation to some provision of the labor agreement or personnel regulations.

G. Pledges that employers required workers to sign indicating that they would not join a union as long as they were employed with the company.

H. A deadlock in negotiations.

I. When employees and employers determine the conditions of employment through direct negotiation.

J. A specific series of steps that must be followed to resolve an employee's dispute, claim, or complaint with labor agreement provisions.

Multiple Choice

Read each item carefully and then select the best response.

_____ **1.** A less powerful form of a written agreement that is often used instead of a labor contract is a:
   A. collective bargaining agreement.
   B. memorandum of understanding.
   C. joint regulation.
   D. good faith agreement.

_____ 2. The nature of the relationship between the employer and the labor organization is determined by a wide variety of labor laws and regulations at the:
  A. federal level.
  B. state level.
  C. local level.
  D. federal, state, and local levels.

_____ 3. Federal legislation provides a set of guidelines for how each state can:
  A. govern public service employees.
  B. negotiate with unions.
  C. regulate collective bargaining.
  D. supervise unionized employees.

_____ 4. An employee who cannot be forced into a contract by an employer in order to obtain and keep a job is covered by the:
  A. Norris-LaGuardia Act of 1932.
  B. Wagner-Connery Act of 1935.
  C. Taft-Hartley Labor Act of 1947.
  D. Landrum-Griffin Act of 1959.

_____ 5. Legislation against yellow dog contracts occurred in the:
  A. Norris-LaGuardia Act of 1932.
  B. Wagner-Connery Act of 1935.
  C. Taft-Hartley Labor Act of 1947.
  D. Landrum-Griffin Act of 1959.

_____ 6. Workers were provided the right to refrain from joining a union in the:
  A. Norris-LaGuardia Act of 1932.
  B. Wagner-Connery Act of 1935.
  C. Taft-Hartley Labor Act of 1947.
  D. Landrum-Griffin Act of 1959.

_____ 7. Collective bargaining was established through the:
  A. Norris-LaGuardia Act of 1932.
  B. Wagner-Connery Act of 1935.
  C. Taft-Hartley Labor Act of 1947.
  D. Landrum-Griffin Act of 1959.

_____ 8. Good faith bargaining and the ability of the president of the United States to apply pressure to achieve resolutions in strike or union discussions is legislated through the:
  A. Norris-LaGuardia Act of 1932.
  B. Wagner-Connery Act of 1935.
  C. Taft-Hartley Labor Act of 1947.
  D. Landrum-Griffin Act of 1959.

_____ 9. The National Labor Relations Board was established by the:
  A. Norris-LaGuardia Act of 1932.
  B. Wagner-Connery Act of 1935.
  C. Taft-Hartley Labor Act of 1947.
  D. Landrum-Griffin Act of 1959.

_____ **10.** The Labor-Management Reporting and Disclosure Act is also known as the:
   A. Norris-LaGuardia Act of 1932.
   B. Wagner-Connery Act of 1935.
   C. Taft-Hartley Labor Act of 1947.
   D. Landrum-Griffin Act of 1959.

_____ **11.** The first documented paid fire department in the United States was the:
   A. Fire Department of New York.
   B. Chicago Fire Department.
   C. Cincinnati Fire Department.
   D. Philadelphia Fire Department.

_____ **12.** FIREPAC promotes the legislative and political interests of the:
   A. IAFF.
   B. NBFSPQ.
   C. IAFC.
   D. NFPA.

_____ **13.** The Fire Service Leadership Partnership was developed in 1999 to address the needs of today's:
   A. fire fighters.
   B. emergency service personnel.
   C. labor–management teams.
   D. fire chiefs and union presidents.

_____ **14.** To ensure that grievances will not be stalled at any level for an excessive time period, there are:
   A. penalties.
   B. timelines.
   C. third-party observers.
   D. peer support teams.

_____ **15.** When dealing with grievances, the objective should be to deal with the problem:
   A. at the lowest possible level.
   B. at the union level.
   C. as the administration deems necessary.
   D. on a one-on-one basis.

_____ **16.** Most grievance procedures are carried out in:
   A. 2 steps.
   B. 3 steps.
   C. 4 steps.
   D. 5 steps.

Fill-in

Read each item carefully, and then complete the statement by filling in the missing word(s).

1. Collective bargaining rights for public employees have traditionally _____ behind those in the private sector.

2. The potential impact on public safety is so severe that many states _____ fire fighters from walking out.

3. A philosophical shift in labor–management relationships is moving from confrontational strategies to _____ relationships.

4. Federal labor laws establish a basic framework that applies to _____ workers.

5. The right-to-work laws were introduced to prohibit the practice of _____ shops.

6. The _____ represents 287,000 fire fighters and emergency service personnel in the United States and Canada.

7. The root cause of almost every labor disturbance is a failure to properly manage the relationship between

   _____ and _____.

8. Extreme negative public reaction and the lasting legacy of lost trust has caused fire fighter strikes to be viewed as

   _____.

9. The managers and supervisors represent the _____ and the union represents the

   _____.

10. The union representative acts as a(n) _____ for the individual or group that submitted the grievance.

Short Answer

Complete this section with short written answers using the space provided.

1. List the steps and procedure for dealing with grievances in your department.

   _____

   _____

   _____

   _____

2. What activities does your union have to assist in healthy labor–management relations?

   _____

   _____

   _____

   _____

3. Identify the key people in your union and labor–management teams.

   _____

   _____

   _____

   _____

4. Identify and discuss any policies, legislations, or agreements between your department and local government or management.

   _____

   _____

   _____

   _____

5. Identify the two main local collective bargaining items that create tension in your department.

   _____

   _____

   _____

   _____

Fire Alarms

The following case studies will give you an opportunity to explore the concerns associated with becoming a fire officer. Read each case study and then answer each question in detail.

1. The fire chief recently put out a memo stating that, "…each shift will start at 0800 instead of 0730 beginning in two weeks." You have been made aware of the policy change by the posted memo. You are approached by several fire fighters who question the change in policy. They advise you that this is a change in working conditions and, by contract, must be negotiated between labor and management. They further advise you that they will be filing an unfair labor practice and grievance with you by the end of the day. How should you proceed?

_____

_____

_____

_____

_____

_____

2. In your organization, the fire department has implemented a rule-by-objective (RBO) approach to labor–management relations. The department is currently looking at reorganizing its volunteer program and wants to provide a 5:1 volunteer-to-volunteer officer ratio. This has created some issues within the career ranks since their supervisory oversight is much higher at a 10:1 ratio. The career union members feel that if the department is going to provide this standard for the volunteers, then the same should apply to the career side of the department to provide a standardized ratio. Further, the career union members believe that making more volunteer lieutenants is going to minimize the career lieutenant's supervisory role in the department and want the department to look at adding a captain rank. The department is not interested in applying the same volunteer-to-volunteer officer ratio to the career ranks and says there are not fiscal resources available to provide a captain in the department. Looking at the RBO method, how would you proceed?

_____

_____

_____

_____

_____

_____

In-Basket

The following exercise will give you an opportunity to refine your writing skills.

1. A fire fighter under your supervision has filed a grievance against another fire fighter on the same shift for inappropriate comments and behavior. Prepare and complete documentation for the grievance and for how you would handle the situation to maintain shift operations.

_____

_____

_____

_____

_____

_____

Workbook Activities

The following activities have been designed to help you. Your instructor may require you to complete some or all of these activities as a regular part of your fire officer training program. You are encouraged to complete any activity that your instructor does not assign as a way to enhance your learning in the classroom.

Chapter Review

The following exercises provide an opportunity to refresh your knowledge of this chapter.

Matching

Match each of the terms in the left column to the appropriate definition in the right column.

_____ **1.** IDLH

A. A hazard and situation assessment.

_____ **2.** Risk management

B. Used to track the identity, assignment, and location of all fire fighters operating at the incident scene.

_____ **3.** Accident

C. The organizational structure used to manage assigned resources in order to accomplish the incident objectives.

_____ **4.** Pre-incident plan

D. Any condition that would pose an immediate or delayed threat to life.

_____ **5.** Incident safety plan

E. The objectives reflecting the overall incident strategy, tactics, risk management, and safety developed by the incident commander.

_____ **6.** Health and safety officer

F. An arrangement of materials and heat sources that presents the potential for harm.

_____ **7.** Rehabilitation

G. The fire department member who manages and investigates all incidents relating to the safety and health program.

_____ **8.** Risk/benefit analysis

H. A written document that provides information that can be used to determine the appropriate actions for an emergency at a specific facility.

_____ **9.** Incident action plan

I. The strategies and tactics developed by the incident safety officer.

_____ **10.** Hazards

J. An unplanned event that interrupts an activity and sometimes causes injury or damage.

_____ **11.** Incident command system

K. Identification and analysis of exposure to hazards and monitoring with respect to the health and safety of members.

_____ **12.** Personnel accountability system

L. The process of providing rest, rehydration, nourishment, and medical evaluation to fire fighters who are involved in incident scene operations.

Multiple Choice

Read each item carefully and then select the best response.

_____ 1. Since 1993, the leading cause of fire fighter fatalities has been:
   A. burns.
   B. trauma.
   C. asphyxiation.
   D. heart attacks.

_____ 2. The fundamental responsibility of every fire officer is to always consider as a primary objective:
   A. knowledge.
   B. motivation.
   C. accountability.
   D. safety.

_____ 3. Most of the traumatic injuries leading to fire fighter fatalities are caused by:
   A. vehicle accidents.
   B. structural collapses.
   C. equipment failure.
   D. burns.

_____ 4. A slogan that should be a constant reminder of safety during emergency incidents is:
   A. "two-in, two-out."
   B. "all for one."
   C. "everyone goes home."
   D. "protect yourself."

_____ 5. The risk of a heart attack is closely related to age and:
   A. race.
   B. sex.
   C. physical fitness.
   D. diet.

_____ 6. The Standard on Health-Related Fitness Programs for Fire Department Members is NFPA:
   A. 1001.
   B. 1021.
   C. 1035.
   D. 1583.

_____ 7. To assist in setting a fitness test specific for fire fighters and their duties, the International Association of Fire Chiefs partnered with the IAFF to develop the:
   A. Cadet Physical Excellence Standard.
   B. Candidate Physical Aptitude Test.
   C. Fire Fighters' Physical Ability Challenge.
   D. Fire Fighters' Fitness Standard.

_____ 8. When driving a fire apparatus, drivers need to be aware that stopping distances are directly related to the vehicle's:
   A. size.
   B. contents.
   C. weight.
   D. wheel base.

_____ **9.** Federal workplace safety regulations in the United States are established by the:
  A. Occupation Safety and Health Administration (OSHA).
  B. National Institute for Occupational Safety and Health (NIOSH).
  C. National Safe Workplace Association (NSWA).
  D. United States Safe Workplace Congress (USSWC).

_____ **10.** The two-in, two-out rule is directly related to the concept of:
  A. scene safety.
  B. teamwork.
  C. the incident management system.
  D. a rapid intervention crew.

_____ **11.** A standardized incident scene process used to ensure personnel do not go unaccounted for is a(n):
  A. incident management system.
  B. personnel accountability system.
  C. rapid intervention crew.
  D. team logistics system.

_____ **12.** The length of time an SCBA provides a safe and reliable air supply depends on the:
  A. level of emergency incident.
  B. equipment maintenance.
  C. user.
  D. IDLH atmosphere.

_____ **13.** During a structure fire, the minimum size of an interior work team is:
  A. two fire fighters.
  B. three fire fighters.
  C. two fire fighters and one officer.
  D. three fire fighters and one officer.

_____ **14.** The incident management system is used by the incident commander to assist in:
  A. managing resources.
  B. coordinating fire fighters.
  C. maintaining accountability.
  D. contacting medical personnel.

_____ **15.** Besides the incident commander, the only officer who has the authority to immediately suspend activities during an imminent hazardous situation is the:
  A. medical director.
  B. operations officer.
  C. fire chief.
  D. safety officer.

_____ **16.** The tactical level management unit that provides for medical evaluation, treatment, nourishment replenishment, and mental rest is the:
  A. rehabilitation center.
  B. scene safety.
  C. incident scene rehabilitation.
  D. incident command center.

_____ **17.** The front line in ensuring compliance with all safety policies is the:
  A. fire chief.
  B. fire officer.
  C. safety officer.
  D. fire fighter.

_____ **18.** The initial investigation of most minor accidents is usually conducted by the:
  A. fire officer.
  B. lieutenant.
  C. deputy chief.
  D. health and safety officer.

Fill-in

Read each item carefully, and then complete the statement by filling in the missing word(s).

1. _____ practices must be the only acceptable behavior, and good _____ habits should be incorporated into all activities.

2. _____ depends on our ability to break the sequence of events that leads to injuries and deaths.

3. For every fire fighter fatality there are approximately _____ fire fighter injuries.

4. Protective clothing and SCBA are designed to _____ the risk of injury.

5. During emergency incidents, fire fighters must work in _____ and fire officers must maintain _____ of all members.

6. It is mandatory that fire fighters use _____ while responding in an emergency vehicle.

7. During an emergency incident, the fire officer must always be prepared for _____ conditions and _____ hazards.

8. An accountability system must be able to function at _____ levels at an incident scene.

Short Answer

Complete this section with short written answers using the space provided.

1. What steps can you take to support and improve fire fighter fitness in your department?

_____

_____

_____

_____

2. Besides protective clothing and equipment, identify the equipment required for fire fighters to carry during emergency incidents.

_____

_____

_____

_____

3. Identify the fire station safety workplace guidelines in your station during normal day-to-day operations.

_____

_____

_____

_____

4. Identify the standard operating procedures used in your organization for infection control or decontamination of equipment, clothing, and personnel.

_____

_____

_____

_____

5. Describe the steps used in accident investigation in your department.

_____

_____

_____

_____

6. List and provide examples of the documentation formats used in your department.

_____

_____

_____

_____

Fire Alarms

The following case studies will give you an opportunity to explore the concerns associated with becoming a fire officer. Read each case study and then answer each question in detail.

1. By using some of the concepts of "everybody goes home," what are some methods that fire officers can use to instill safe work practices into the everyday routine of the shift and ensure that no preventable accidents or deaths occur on duty?

_____

_____

_____

_____

_____

2. While responding to a basic life support call, a vehicle pulls out in front of the fire engine and a minor traffic accident occurs. No one is injured and you advise dispatch of the situation and ask for another unit to be dispatched. You request that a chief officer respond to the scene and begin to complete the accident report form found on the apparatus. The individual in the car states that he or she never saw the fire engine responding. You clear the scene and return to quarters to determine the damage done to the apparatus. What else needs to occur?

_____

_____

_____

_____

_____

In-Basket

The following exercises will give you an opportunity to refine your writing skills.

1. Using your shift schedule, identify exercise/workout times for the next three weeks.

_____

_____

_____

_____

_____

_____

2. Provide complete documentation, including investigation notes, for a work-related accident or hazardous exposure.

_____

_____

_____

_____

_____

3. Prepare a training presentation for your company promoting the use of seat belts on all fire apparatus.

_____

_____

_____

_____

_____

Training and Coaching

Workbook Activities

The following activities have been designed to help you. Your instructor may require you to complete some or all of these activities as a regular part of your fire officer training program. You are encouraged to complete any activity that your instructor does not assign as a way to enhance your learning in the classroom.

Chapter Review

The following exercises provide an opportunity to refresh your knowledge of this chapter.

Matching

Match each of the terms in the left column to the appropriate definition in the right column.

_____ 1. Education   **A.** The process of achieving proficiency through instruction and hands-on practice in the operation of equipment and systems.

_____ 2. Accreditation   **B.** To provide training to an individual or a team.

_____ 3. Pro Board   **C.** Standard for Fire Fighter Professional Qualifications.

_____ 4. NFPA 1021   **D.** Standard for Fire Department Safety Officer.

_____ 5. Coach   **E.** A systematic four-step approach to training fire fighters in a basic job skill.

_____ 6. NFPA 1001   **F.** A system used by a certification organization that determines a school or program meets with the requirements of the fire service.

_____ 7. NFPA 1041   **G.** The process of imparting knowledge or skill through systematic instruction.

_____ 8. Training   **H.** National Professional Qualifications System.

_____ 9. Job instruction training   **I.** Standard for Fire Officer Professional Qualifications.

_____ 10. NFPA 1521   **J.** Standard for Fire Service Instructor Professional Qualifications.

Multiple Choice

Read each item carefully and then select the best response.

_____ 1. The first organized instruction of fire officers in 1869 was in:
- **A.** Chicago.
- **B.** Los Angeles.
- **C.** New York.
- **D.** Cincinnati.

_____ **2.** The National Association of State Directors of Fire Training established the:
   A. National Professional Qualifications System.
   B. International Fire Service Accreditation Congress.
   C. National Fire Protection Association.
   D. Fire and Emergency Services Higher Education conference.

_____ **3.** Certification usually involves formal training that includes:
   A. the latest technologies.
   B. new equipment.
   C. all emergency service personnel.
   D. classroom and skills practice.

_____ **4.** The four-step method of skill training follows:
   A. discuss, demonstrate, do, and diagnose.
   B. introduce, overview, practice, and review.
   C. prepare, present, apply, and evaluate.
   D. prepare, philosophize, practice, and assess.

_____ **5.** The first step of the four-step method of instruction focuses on:
   A. the information or skill.
   B. the students.
   C. the training environment.
   D. instructor readiness.

_____ **6.** The lecture or instructional portion of the training occurs in the:
   A. overview step.
   B. preparation step.
   C. discussion step.
   D. presentation step.

_____ **7.** To assist and allow the fire officer to stay on topic and emphasize the important points, he or she should use a:
   A. lesson plan.
   B. daily log.
   C. visual aid.
   D. textbook.

_____ **8.** To assist fire fighters in retaining information more effectively, they should:
   A. be formally evaluated.
   B. have open discussions.
   C. perform the skills.
   D. do homework assignments.

_____ **9.** Depending on the skill and knowledge sets being trained, evaluations may be:
   A. informal.
   B. written or practical.
   C. completed as a group.
   D. unnecessary.

_____ **10.** To be certain training has occurred correctly, the fire fighter's performance must show:
    A. an observable change.
    B. complete mastery of the skills.
    C. the new skill in practice.
    D. that he or she can discuss the new information at an advanced level.

_____ **11.** The emphasis of fire station–based training should be on:
    A. the development of new skills.
    B. safety and effective use of the device or procedure.
    C. the use of new equipment.
    D. teamwork.

_____ **12.** The ability of a driver/operator to remember an area of the district he has not had to respond to in several months is an example of the:
    A. initial level of psychomotor skills.
    B. plateau level of psychomotor skills.
    C. latency level of psychomotor skills.
    D. mastery level of psychomotor skills.

_____ **13.** For fire fighters to have confidence in seldom-used skill sets, they must:
    A. work in teams.
    B. have enough repetitions and simulations with the skills.
    C. be regularly evaluated to NFPA standards.
    D. be in the plateau level of psychomotor skills.

_____ **14.** Bloodborne pathogens, hazardous materials awareness, and SCBA fit testing are training topics that fire fighters are required to take by:
    A. federal regulations.
    B. state or provincial legislation.
    C. local or municipal collective bargaining agreement.
    D. department standard operation procedures.

_____ **15.** Before a fire officer develops a training program, he or she should:
    A. identify the skills to train.
    B. assess needs.
    C. form objectives.
    D. develop the evaluation process.

_____ **16.** A lesson plan outlines:
    A. what to teach.
    B. what order to teach.
    C. what procedures to follow.
    D. All of the above.

Fill-in

Read each item carefully, and then complete the statement by filling in the missing word(s).

1. The National Fire Protection Association started a trend of developing national _____ standards on a wide range of fire service occupations.

2. The National Fire Protection Association standards define the _____ that an individual must demonstrate to be certified at a given level.

3. Accreditation establishes the _____ of the system to award _____ that are based on the standards.

4. An indicator that training is needed would be an observed performance _____.

5. When instructing, begin with _____ concepts and progress to more _____ information, relating new information to old ideas.

6. To assist learning, the fire officer should provide immediate _____ to identify omissions and correct errors.

7. It is important that some practice sessions be performed while simulating _____ fire-ground situations.

8. During training the fire officer should spend time developing _____ and relating the training to _____ procedures.

9. Responding to alarms, on-scene activity, and emergency procedures are three behaviors the fire officer must spell out directly for the _____.

10. NFPA 1403 specifies that during live fire evolutions, no _____ "victims" or _____ liquids may be used and that only _____ fire evolution(s) may be run in an acquired structure at a time.

Short Answer

Complete this section with short written answers using the space provided.

1. Outline and describe the sections of the lesson plan format utilized in your department.

_____

_____

_____

_____

2. Identify the expected behaviors of new fire fighters and trainees in your department during emergency responses.

_____

_____

_____

_____

3. Include the training action plan for live fire training evolutions utilized in your department.

_____

_____

_____

_____

4. Identify safety precautions taken during live fire training evolutions in your department, as well as the procedure for dealing with an injured fire fighter during a training activity.

_____

_____

_____

_____

5. Identify two of your strengths and challenges as an instructor.

_____

_____

_____

_____

Fire Alarms

The following case studies will give you an opportunity to explore the concerns associated with becoming a fire officer. Read each case study and then answer each question in detail.

1. While on shift, you are called into the battalion chief's office and asked to help put together a class on the use of extrication equipment. The battalion chief would like you to teach this course to the entire department on a shift basis. He is requesting a formal class with outline, handouts, and a practical skills session. The training will be scheduled in two weeks on three consecutive days. How will you proceed in putting together this course?

_____

_____

_____

_____

_____

_____

2. As the fire officer in charge of training, as a part of your strategic planning process, you have been tasked with getting your agency's Fire Fighter I training program accredited. You contact other departments in the area and none of them currently have accredited programs nor are they familiar with such a process. What does accreditation mean and what are the options available?

_____

_____

_____

_____

_____

In-Basket

The following exercises will give you an opportunity to refine your writing skills.

1. Develop a specific training program outline, with lesson plans, for new fire fighter and/or trainee orientation.

   _____

   _____

   _____

   _____

   _____

   _____

2. Identify a training need in your department and prepare a report outlining the resources required, scheduling considerations, and the process for evaluation or certification of training.

   _____

   _____

   _____

   _____

   _____

   _____

3. Prepare a lesson plan for a presentation promoting the use of seat belts on all fire apparatus. Include your presentation objectives and a student evaluation.

   _____

   _____

   _____

   _____

   _____

   _____

Evaluation and Discipline

Workbook Activities

The following activities have been designed to help you. Your instructor may require you to complete some or all of these activities as a regular part of your fire officer training program. You are encouraged to complete any activity that your instructor does not assign as a way to enhance your learning in the classroom.

Chapter Review

The following exercises provide an opportunity to refresh your knowledge of this chapter.

Matching

Match each of the terms in the left column to the appropriate definition in the right column.

_____ 1. Involuntary transfer

_____ 2. Discipline

_____ 3. Restrictive duty

_____ 4. Suspension

_____ 5. Central tendency

_____ 6. Recency

_____ 7. Termination

_____ 8. Oral reprimand

_____ 9. Personal bias

_____ 10. Demotion

A. The first level of negative discipline.

B. The organization has determined that the member is unsuitable for continued employment.

C. A reduction in rank, with a corresponding reduction in pay.

D. A negative disciplinary action that removes a fire fighter from a work location.

E. An evaluation error that occurs when an evaluator allows his or her personal feelings to affect a fire fighter's evaluation.

F. A disciplinary action where a fire fighter is transferred to a less desirable work location.

G. An evaluation error that occurs when a fire fighter is rated only on recent incidents rather than the entire evaluation period.

H. An evaluation error that occurs when a fire fighter is rated in the middle range for all dimensions of work performance.

I. A moral, mental, and physical state in which all ranks respond to the will of the leader.

J. A temporary work assignment that isolates the fire fighter from the public.

Multiple Choice

Read each item carefully and then select the best response.

_____ 1. Supervision and feedback that is intended to help the employee recognize problems and make corrections to improve performance or behavior is:
   A. proactive discipline.
   B. reactive discipline.
   C. progressive discipline.
   D. positive discipline.

_____ **2.** Supervision and feedback that is punishment for unsatisfactory performance and/or unacceptable behavior is:
   A. directed discipline.
   B. corrective discipline.
   C. negative discipline.
   D. punitive discipline.

_____ **3.** Most career fire departments conduct formal:
   A. monthly performance evaluations.
   B. bimonthly performance evaluations.
   C. semiannual performance evaluations.
   D. annual performance evaluations.

_____ **4.** By providing a comprehensive and effective probationary period, fire officers have a special opportunity to:
   A. prepare future departmental leaders.
   B. develop new training methods.
   C. observe the operational abilities of the department.
   D. interact with all levels of emergency personnel.

_____ **5.** In general, the subordinate is allowed to review and comment on his or her performance evaluations in the:
   A. first step of the annual evaluation process.
   B. second step of the annual evaluation process.
   C. third step of the annual evaluation process.
   D. exit interview.

_____ **6.** The final step of an annual evaluation should:
   A. include a human resources professional.
   B. have union representation.
   C. be conducted by the fire chief.
   D. establish goals for the next evaluation period.

_____ **7.** To increase focus on improving performance and achievement for the next evaluation period, fire fighters should:
   A. identify work-related goals.
   B. document daily activities.
   C. focus on teamwork.
   D. take computer training.

_____ **8.** A formal review session to assess progress that includes a fire fighter self-evaluation is a:
   A. progress report.
   B. mid-year review.
   C. personal critique meeting.
   D. self-assessment.

_____ **9.** If a fire fighter is evaluated on the basis of the fire officer's personal ideals instead of the classified job standards, there is a:
   A. personal bias evaluation error.
   B. severity evaluation error.
   C. frame of reference evaluation error.
   D. central tendency evaluation error.

_____ **10.** If a fire officer compares the performance of one fire fighter with another fire fighter, there is a:
A. haloing effect evaluation error.
B. contrast effect evaluation error.
C. comparative evaluation error.
D. peer biasing evaluation error.

_____ **11.** As a general rule, positive discipline should always be used:
A. after negative discipline.
B. prior to annual evaluations.
C. in conjunction with empowerment strategies.
D. before negative discipline.

_____ **12.** Convincing someone he or she wants to do better and is capable and willing to make the effort is the key to:
A. positive discipline.
B. annual evaluations.
C. providing feedback.
D. decreasing the severity of a reprimand.

_____ **13.** The starting point for positive discipline is to establish:
A. a set of expectations for behavior and performance.
B. good documentation records.
C. a formal meeting.
D. an observation schedule.

_____ **14.** Fire fighters can gauge a fire officer's level of commitment by observing the characteristics of:
A. positive attitude.
B. self discipline.
C. teamwork.
D. high standards.

_____ **15.** The disciplinary process is designed to be:
A. subjective and personalized.
B. clear and concise.
C. consistent and well documented.
D. formalized and open.

_____ **16.** In most organizations, the third step of the negative discipline process is a(n):
A. informal written reprimand.
B. formal written reprimand.
C. suspension.
D. termination.

_____ **17.** Prior to a suspension, demotion, or involuntary termination, there is often a(n):
A. predisciplinary conference.
B. media involvement.
C. offer of financial restitution.
D. loss of holidays or benefits.

_____ **18.** Personnel actions, grievances, evaluation reports, and hiring packet information is secured in the:
A. fire chief's records.
B. employee assistance program.
C. fire officer's documentation.
D. employee's personnel file in a secure central repository.

Fill-in

Read each item carefully, and then complete the statement by filling in the missing word(s).

1. Fire officers are required to provide regular evaluations to give fire fighters feedback on _____, _____, and _____.

2. The ultimate level of negative discipline for an employee is _____.

3. If the evaluation process is performed and handled correctly, it should assist the employee by providing personal _____ and motivation.

4. In career fire departments, there is usually a classified _____ that specifies all of the required knowledge, skills, and abilities that a fire fighter is expected to master to complete the probationary requirements.

5. In general, there are _____ steps in an annual evaluation.

6. For the fire officer to provide an objective, fair, and complete annual performance evaluation, it is essential to have _____ an employee's performance and progress throughout the year.

7. A(n) _____ is a documentation system similar to an accounting balance sheet listing assets on the left side and liabilities on the right side of the sheet.

8. A fire fighter who is not meeting expectations should know there is a problem _____ the annual evaluation.

9. Employees who receive a substandard annual evaluation may receive a work improvement plan that should include a specific _____ for expected improvements.

10. During face-to-face meetings, fire officers may be _____ to reduce conflict.

Short Answer

Complete this section with short written answers using the space provided.

1. Identify three external and internal factors that may affect a fire fighter's work performance.

_____
_____
_____
_____

2. Provide and describe a copy of the performance evaluation used in your department.

_____
_____
_____
_____

3. Identify three special considerations you would incorporate in a probationary or new fire fighter's evaluation versus a veteran department member's evaluation.

_____
_____
_____
_____

4. How can you empower your fire fighters in their performance of day-to-day activities?

_____

_____

_____

_____

5. Outline the negative discipline process in your department.

_____

_____

_____

_____

6. Identify the predetermined disciplinary policies supported in your department.

_____

_____

_____

_____

Fire Alarms

The following case studies will give you an opportunity to explore the concerns associated with becoming a fire officer. Read each case study and then answer each question in detail.

1. Fire Fighter Bradbury is due for his annual performance evaluation and Lieutenant Novak is beginning to work on the process. Going through the performance evaluation form, Lieutenant Novak poses the question, "What positive strengths has this fire fighter contributed to the fire company?" He knows that Fire Fighter Bradbury has been a good member of the shift, but he has difficulty in being able to remember and cite specific instances of his positive contributions and strengths. What would make it easier for Lieutenant Novak to complete his evaluation and be able to identify past work performance?

_____

_____

_____

_____

_____

_____

2. Assistant Chief Henderson is in charge of operations and training, and as part of those duties he is responsible for ensuring that each shift is meeting the training requirements of the department. As part of this responsibility, Assistant Chief Henderson developed a new skills evaluation process for each shift. On a quarterly basis, each shift is evaluated by the assistant chief on an evolution of his choice. Today, he is conducting the skills evaluation for "A" shift on a forward lay evolution with quick attack. The evolution does not go as planned and it is evident that the shift has not been doing the prescribed training. What would be an appropriate action for Assistant Chief Henderson to take?

_____

_____

_____

_____

_____

In-Basket

The following exercises will give you an opportunity to refine your writing skills.

1. Write a mid-year review self-evaluation, readdressing your annual evaluation work-related goals, for a potential meeting with your supervisor.

_____

_____

_____

_____

_____

_____

2. Write a written reprimand for a fire fighter who repeatedly arrives at work late.

_____

_____

_____

_____

_____

3. Prepare a written reprimand for a company member who slept through an emergency run.

_____

_____

_____

_____

_____

Leading the Fire Company

Workbook Activities

The following activities have been designed to help you. Your instructor may require you to complete some or all of these activities as a regular part of your fire officer training program. You are encouraged to complete any activity that your instructor does not assign as a way to enhance your learning in the classroom.

Chapter Review

The following exercises provide an opportunity to refresh your knowledge of this chapter.

Matching

Match each of the terms in the left column to the appropriate definition in the right column.

_____ **1.** Laissez-faire   **A.** Motivational theory that identifies internal and external factors.

_____ **2.** Expectancy theory   **B.** The leadership approach required in a high-risk, emergency scene activity.

_____ **3.** Leadership   **C.** Leaders can be effective only to the extent that followers are willing to accept their leadership.

_____ **4.** Motivation   **D.** When the decision making moves from the leader to the followers.

_____ **5.** Democratic   **E.** When people act in a manner that they believe will lead to an outcome they value.

_____ **6.** Extinction   **F.** The consultative approach to making decisions.

_____ **7.** Autocratic   **G.** Inspiring others to achieve their maximum potential.

_____ **8.** Negative reinforcement   **H.** When bad behavior is ignored.

_____ **9.** Motivation-hygiene theory   **I.** Removing an undesirable consequence of good behavior.

_____ **10.** Followership   **J.** When a person directs an organization in a way that makes it more cohesive and coherent.

Multiple Choice

Read each item carefully and then select the best response.

_____ **1.** The theory that suggests that behavior is a function of its consequences is:
   A. reinforcement theory.
   B. motivation-hygiene theory.
   C. goal-setting theory.
   D. equity theory.

2. When employees evaluate the outcomes that they receive for their inputs and compare them with the outcomes others receive for their inputs is described by:
   A. reinforcement theory.
   B. motivation-hygiene theory.
   C. goal-setting theory.
   D. equity theory.

3. According to reinforcement theory, the reinforcer that is often underutilized is:
   A. positive reinforcement.
   B. corrective reinforcement.
   C. negative reinforcement.
   D. supportive reinforcement.

4. The leadership style that utilizes the resourcefulness and strengths of the group is the:
   A. autocratic style.
   B. laissez-faire style.
   C. dictator style.
   D. democratic style.

5. The leadership style that allows followers to utilize their judgment and sense of responsibility to accomplish a task is the:
   A. autocratic style.
   B. laissez-faire style.
   C. dictator style.
   D. democratic style.

6. When immediate action is required, the fire officer should utilize the:
   A. autocratic leadership style.
   B. laissez-faire leadership style.
   C. dictator leadership style.
   D. democratic leadership style.

7. One of the responsibilities of the first-arriving fire company at an incident scene is to:
   A. estimate the amount of damage.
   B. initiate an interior attack.
   C. describe situations to the dispatch center.
   D. determine the cause of the incident.

8. The first-arriving fire officer implements the:
   A. incident management plan.
   B. offensive attack orders.
   C. defensive attack orders.
   D. department code of response.

9. The two most important values that every fire fighter needs to respect are responding to emergencies and:
   A. saving property.
   B. preserving the public trust.
   C. working as a team.
   D. the efficient use of resources.

_____ **10.** During routine activities, today's effective fire officers provide a leadership style that is more:
  A. autocratic.
  B. individualized.
  C. participative.
  D. informal.

_____ **11.** According to reinforcement theory, using punishment and extinction:
  A. increases the likelihood of bad behavior.
  B. decreases the likelihood of good behavior.
  C. assists in establishing goals.
  D. does not guarantee good behavior.

_____ **12.** Historically, in both emergency and nonemergency duties, fire officers used a(n):
  A. autocratic leadership style.
  B. laissez-faire leadership style.
  C. dictator leadership style.
  D. democratic leadership style.

_____ **13.** A core responsibility of a fire officer is to:
  A. motivate fire fighters.
  B. handle emergencies effectively.
  C. administer fire department budgets.
  D. hire effective fire fighters.

_____ **14.** One of the key components of leading is the ability to:
  A. achieve.
  B. administer.
  C. follow.
  D. motivate.

_____ **15.** Radio reports to the dispatch center should be:
  A. documented by the incident commander.
  B. made after the initial attack plans are made.
  C. calm, concise, and complete.
  D. used to assist in developing an action plan.

_____ **16.** Often the strongest force that influences volunteer fire fighter performance and commitment is:
  A. compensation or benefits.
  B. opportunities for promotion.
  C. effective leadership.
  D. new equipment or uniforms.

Fill-in

Read each item carefully, and then complete the statement by filling in the missing word(s).

1. The key to motivation when using the goal-setting theory is to set _____ goals that are difficult, but _____.

2. A fire officer always has direct leadership _____ for the company that he or she is commanding.

3. According to reinforcement theory, positive and negative reinforcement _____ the likelihood of good behavior.

4. Bosses tell people to accomplish an objective, whereas leaders make them _____ to achieve the objective.

5. Most people agree that an effective leader changes the style of leadership based on the specific _____.

6. Effective leaders are also good _____, supporting the goals and objectives of their leaders and department.

7. An employee's belief that his or her effort will achieve a goal is a motivational consideration of _____ theory.

8. According to Herzberg's theory, to determine the fire fighter's needs the fire officer must have open and honest _____ with the individual fire fighters.

9. Hygiene factors are _____ to the individual, whereas the motivation factors are _____ to the individual.

10. A goal of an effective fire officer is to push decision making to the _____ possible level.

Short Answer

Complete this section with short written answers using the space provided.

1. Which of your personal characteristics will assist you in establishing yourself as a leader?

_____

_____

_____

_____

2. How does your fire department assign tasks on scene?

_____

_____

_____

_____

3. Outline your fire department's operating procedures for the initial radio report when arriving on scene.

_____

_____

_____

_____

4. Identify your preferred leadership style.

_____

_____

_____

_____

5. How would you assist in motivating an underachieving fire fighter?

_____

_____

_____

_____

6. What steps could you take to assist in motivating a crew of fire fighters?

_____

_____

_____

_____

Fire Alarms

The following case studies will give you an opportunity to explore the concerns associated with becoming a fire officer. Read each case study and then answer each question in detail.

1. As a new fire officer, you decide that it would be beneficial to sit down one-on-one with each member of your company in order to get to know them better and to determine what their needs are. You call in Fire Fighter Link, the most senior member of your company, and have an open dialogue with him. He is forthcoming, and after the meeting thanks you for listening, and for the opportunity to meet with you. In addition to one-on-one meetings with company members, what are some ways that company officers can receive both positive and negative feedback from company members? What are some useful ways to learn from any criticism?

_____

_____

_____

_____

_____

_____

2. Your fire company is dispatched to a residential structure fire. Prior to the alarm, your company has been conducting hose drills at the nearby elementary school and just completed an evolution. Your company quickly picks up the hose and equipment and responds to the fire. As you approach, a smoke column is visible and you know this will be a working fire. You work to get all your equipment back to normal en route to the call and are frustrated when everything is not where it should be. When you arrive, your frustration lends itself to an incomplete size-up and a disorganized start to the initial attack and setup. The first-arriving company officer is responsible for initial size-up, as well as a host of other things. When a size-up is given in a hurried, confusing, or indecisive manner, what kind of a mood will this set for the incident?

_____

_____

_____

_____

_____

_____

In-Basket

The following exercises will give you an opportunity to refine your writing skills.

1. Develop your leadership statement.

_____
_____
_____
_____
_____
_____

2. Develop an outline of five questions that you could use to interview individual fire fighters to find what motivates them.

_____
_____
_____
_____
_____
_____

3. Using the information and theories in this chapter, evaluate your present form of leadership and identify three changes to improve your current leadership style.

_____
_____
_____
_____
_____
_____

Working in the Community

Workbook Activities

The following activities have been designed to help you. Your instructor may require you to complete some or all of these activities as a regular part of your fire officer training program. You are encouraged to complete any activity that your instructor does not assign as a way to enhance your learning in the classroom.

Chapter Review

The following exercises provide an opportunity to refresh your knowledge of this chapter.

Matching

Match each of the terms in the left column to the appropriate definition in the right column.

_____ 1. Demographics

_____ 2. Press release

_____ 3. Individual needs

_____ 4. Interview

_____ 5. Systematic needs

_____ 6. Public information officer

_____ 7. Risk Watch

_____ 8. Public safety education

_____ 9. CERT

_____ 10. Fire Prevention Week

A. Issues that are addressed at the departmental or community level.

B. The first comprehensive injury prevention program available for use in schools.

C. A program that has the fundamental goal of preventing injury, death, or loss due to fire or other types of incidents.

D. The characteristics of human populations and population segments.

E. A program that helps citizens understand their responsibilities in preparing for disaster.

F. Issues that already have programs in place to address the need.

G. The week in which October 8 falls, commemorating the anniversary of the Great Chicago Fire of October 8, 1871.

H. An official announcement to the news media from the fire department public information officer.

I. The source of official fire department information.

J. The most common method where a fire officer would appear as an official spokesperson.

Multiple Choice

Read each item carefully and then select the best response.

_____ 1. Today the relationship between many fire departments and their communities in hundreds of cities has changed due to:
   A. unpredictable response times.
   B. administrative policies.
   C. standard operating procedures.
   D. urban decay and civil unrest.

_____ **2.** The federal government undertakes a nationwide census once every:
- **A.** 2 years.
- **B.** 5 years.
- **C.** 10 years.
- **D.** 12 years.

_____ **3.** From 1996 to 2003, the working age population grew _____ due to recent immigrants.
- **A.** 30%
- **B.** 40%
- **C.** 50%
- **D.** 60%

_____ **4.** Demographic information allows the fire officer to identify the characteristics of a community before:
- **A.** delivering a safety message.
- **B.** developing standard operating procedures.
- **C.** responding to an incident.
- **D.** promoting fire fighters for officer duties.

_____ **5.** The primary goal of most fire departments is to:
- **A.** limit the damage of emergency incidents.
- **B.** save lives and property.
- **C.** respond to incidents as quickly as possible.
- **D.** prevent fires.

_____ **6.** Fire officers need to be aware of what community programs are available, often through the health and:
- **A.** social services departments.
- **B.** education departments.
- **C.** community awareness departments.
- **D.** municipal departments.

_____ **7.** When a citizen makes a request not within the fire officer's authority, the fire officer should respectfully:
- **A.** deny the request.
- **B.** discuss the issue to the best of his or her ability.
- **C.** listen and document the conversation.
- **D.** provide direction to where the request can be resolved.

_____ **8.** The goals, content, and delivery mechanisms for a public education program vary for fire departments, depending on:
- **A.** the department's level of service.
- **B.** the department's resources and circumstances.
- **C.** management support.
- **D.** the interests of the fire fighters.

_____ **9.** An educational presentation is successful when it causes a(n):
- **A.** emotional response from the audience.
- **B.** feeling of satisfaction.
- **C.** change of behavior.
- **D.** request for that particular service.

_____ **10.** The history of Fire Prevention Week has its roots in the:
   A. Philadelphia Fire Department.
   B. Cincinnati Fire Department.
   C. Peshtigo Fire.
   D. Great Chicago Fire.

_____ **11.** Every year since 1922, National Fire Prevention Week has been observed on the Sunday-through-Saturday period in which:
   A. September 25th falls.
   B. February 12th falls.
   C. March 3rd falls.
   D. October 8th falls.

_____ **12.** The first session of the CERT course in the FEMA version is:
   A. Disaster Preparedness.
   B. Learn Not to Burn.
   C. Stop, Drop, and Roll.
   D. Change Your Clock—Change Your Battery.

_____ **13.** When developing an education program, the developers should follow the:
   A. three-step planning process.
   B. five-step planning process.
   C. seven-step planning process.
   D. nine-step planning process.

_____ **14.** When speaking to the media, never assume:
   A. that your entire interview will be used.
   B. that anything you say is "off the record."
   C. that you will be presented in a positive image.
   D. that the correct information will be presented.

_____ **15.** After answering a reporter's question:
   A. stop talking.
   B. ask him or her to repeat the answer to you.
   C. check for clarification points.
   D. direct the reporter to other fire officers.

_____ **16.** Every fire officer should be prepared to act as a:
   A. fire department spokesperson.
   B. politician.
   C. fire house cook.
   D. demographic expert.

Fill-in

Read each item carefully, and then complete the statement by filling in the missing word(s).

1. Each community has special _____ and has different _____ that should be considered in relation to every service and program provided by the fire department.

2. The best method of avoiding fire injuries and deaths is by _____ the fire.

3. The trend of local government in the early 21st century is to move toward more _____ local government.

4. Cultural _____ is required to help customers of different origins appreciate the services provided by the fire department.

5. Progressive fire departments have moved toward prevention; this type of thinking often requires a shift in the _____ of the fire department.

6. Fire officers are more likely to take action in a fire department that _____ the fire officer to initiate direct action.

7. If a citizen makes a fire officer aware that he or she is contacting the fire administration, the fire officer should _____ the fire administration.

8. In many cases, a public education program is developed at the local level to meet _____ needs.

9. Even if a fire department has a public information officer, every fire officer should be prepared to act as a(n) _____ for the fire department.

10. When talking to a reporter, assume that everything you say could be _____, including conversations before and after the actual interview.

Short Answer

Complete this section with short written answers using the space provided.

1. Describe the demographics of your fire department's community.

_____

_____

_____

_____

2. How do the demographics of your fire department affect your response and service?

_____

_____

_____

_____

3. Identify the public education programs and initiatives actively promoted by your fire department.

_____

_____

_____

_____

4. Identify a public education program, not presently offered by your fire department, that would be beneficial to your community.

_____

_____

_____

_____

5. Identify your fire department's guidelines and training for dealing with the media.

_____

_____

_____

_____

Fire Alarms

The following case studies will give you an opportunity to explore the concerns associated with becoming a fire officer. Read each case study and then answer each question in detail.

1. Your chief has recently informed you at your monthly staff meeting that your fire department is planning to have a fire safety and prevention booth at the upcoming county fair. The chief sees this as a useful public relations tool and has given you the task as a fire officer to organize and plan for the event. What can you do to make this fire safety event informative to the public?

_____

_____

_____

_____

_____

2. Your engine company is dispatched to a vehicle fire on the east side of town. After arriving at the dispatched address, you inform dispatch that you are unable to locate the scene. Dispatch reports that they have additional calls about the car fire being in a completely different location on the opposite side of the district. You turn around and drive to the new location only to find that the vehicle fire is fully involved. Several onlookers are shaking their heads at the delayed response. What can you do to explain to the onlookers the importance of giving accurate information to a dispatcher? How can you use this incident as an example of what can happen if faulty information is relayed to responders?

_____

_____

_____

_____

_____

In-Basket

The following exercises will give you an opportunity to refine your writing skills.

1. Write a media release to promote one of your fire department's public education programs.

   _____

   _____

   _____

   _____

   _____

   _____

2. A local controversial community-based organization has approached the fire department for special fundraising initiatives. Appreciating the potential for political and local concern, write a letter to deny the fire department's participation.

   _____

   _____

   _____

   _____

   _____

   _____

3. Prepare a press release for a total fire loss in an industrial structure where the cause is undetermined, there was no loss of life, 30 employees are now out of work, and the company is fully insured.

   _____

   _____

   _____

   _____

   _____

Handling Problems, Conflicts, and Mistakes

Workbook Activities

The following activities have been designed to help you. Your instructor may require you to complete some or all of these activities as a regular part of your fire officer training program. You are encouraged to complete any activity that your instructor does not assign as a way to enhance your learning in the classroom.

Chapter Review

The following exercises provide an opportunity to refresh your knowledge of this chapter.

Matching

Match each of the terms in the left column to the appropriate definition in the right column.

_____ 1. Investigation	**A.** Information shared by people who have been through a program or experience.
_____ 2. Customer service	**B.** A method of shared problem solving in which all members of a group spontaneously contribute ideas.
_____ 3. Deadlines	**C.** An error or fault resulting from defective judgment, deficient knowledge, or carelessness.
_____ 4. Complaint	**D.** An individual who presents an accusation or statement of regret.
_____ 5. Complainant	**E.** A systematic inquiry or examination.
_____ 6. Conflict	**F.** The conscious process of securing all kinds of information through a combination of listening and observing.
_____ 7. Feedback	**G.** An expression of grief, resentment, accusation, or fault finding.
_____ 8. Mistake	**H.** Useful in focusing effort and prioritizing activities.
_____ 9. Active listening	**I.** A process used to fix problems and provide information to the public.
_____ 10. Brainstorming	**J.** A state of opposition between two parties.

Multiple Choice

Read each item carefully and then select the best response.

_____ 1. Whenever a fire officer is faced with a situation that requires a response, he or she is required to use his or her:
   A. physical abilities.
   B. administrative and department policies.
   C. decision-making skills.
   D. team.

_____ **2.** Problem-solving techniques are designed to identify and evaluate the realistic and potential solutions to a problem and:
- **A.** utilize the best resources available.
- **B.** determine the best decision.
- **C.** efficiently mediate the response.
- **D.** administer department policies.

_____ **3.** A complaint about the assignment of duties to different individuals within a fire station can be identified as a(n):
- **A.** in-house issue.
- **B.** internal issue.
- **C.** external issue.
- **D.** low-profile incident.

_____ **4.** Fire departmental activities that involve private citizens may create:
- **A.** in-house issues.
- **B.** internal issues.
- **C.** external issues.
- **D.** low-profile incidents.

_____ **5.** As the fire officer moves up through the ranks, the number of decision-making situations:
- **A.** decreases.
- **B.** stays the same but becomes more policy based.
- **C.** becomes more external.
- **D.** increases exponentially.

_____ **6.** The first step to solving any problem should be to:
- **A.** define the problem.
- **B.** identify the people involved.
- **C.** refer to the departmental policies.
- **D.** predict the possible outcomes.

_____ **7.** A fire officer should make a conscious effort to identify activities that can be:
- **A.** documented.
- **B.** good training scenarios.
- **C.** changed, improved, or updated.
- **D.** addressed and dealt with by all department members.

_____ **8.** The goal of every fire officer should be to foster a(n):
- **A.** amicable relationship with all employees.
- **B.** trusting relationship with all employees.
- **C.** open environment in the workplace.
- **D.** enjoyable place to work.

_____ **9.** The best people to solve a problem are usually those who:
- **A.** caused the problem.
- **B.** observed the problem.
- **C.** have specialized training in dealing with the problem.
- **D.** are directly involved with the problem.

_____ **10.** Often the presence of a senior officer:
   A. intimidates members of the department.
   B. influences the behavior of a group.
   C. indicates the problem is serious.
   D. shows interest from the administration.

_____ **11.** After a decision has been made, often the most challenging aspect of problem solving is the:
   A. policy development.
   B. evaluation of the results.
   C. implementation phase.
   D. corrective behavior training.

_____ **12.** To assist in clearly assigning tasks, a fire officer should list tasks, responsibilities, and due dates on a(n):
   A. "to-do" list.
   B. project plan.
   C. internal memorandum.
   D. goal sheet.

_____ **13.** For a problem to be truly solved, the solution must be:
   A. implemented.
   B. documented.
   C. observed.
   D. supported by the fire fighters.

_____ **14.** The part of the process used to determine if a solution produced the desired results is the:
   A. evaluation phase.
   B. feedback phase.
   C. follow-up observations.
   D. training review.

_____ **15.** A fire officer becomes officially involved in a problem when the officer:
   A. has to deal with the problem.
   B. addresses the department members.
   C. becomes aware the problem exists.
   D. provides guidance to the complainant.

_____ **16.** During the first phase of the conflict resolution model, the fire officer should:
   A. try to understand the issue.
   B. document all information.
   C. gather all members involved in the problem.
   D. discuss the problem with another officer.

_____ **17.** Once an investigation is completed, the fire officer:
   A. should provide the solution to solve the problem.
   B. determines who to involve in the corrective action(s).
   C. addresses all parties of the investigation.
   D. presents the findings to a supervisor.

_____ **18.** When dealing with a highly emotional situation, the first step is often to:
   A. take control of the scene.
   B. listen actively.
   C. separate any and all individuals involved.
   D. identify the source of the problem.

_____ **19.** Customer satisfaction focuses on:
   A. achieving the fire department's mission statement.
   B. serving the public.
   C. meeting the customer's expectations.
   D. efficiency of service.

_____ **20.** If a proposed solution to a problem involves discipline, the company officer is obligated to:
   A. protect employee privacy.
   B. suspend the fire fighter.
   C. segregate the complainant.
   D. contact the fire department attorney.

Fill-in

Read each item carefully, and then complete the statement by filling in the missing word(s).

1. In general, a(n) _____ is the difference between the actual state and the desired state.

2. One of the most critical areas in decision making is how to deal with situations that involve _____.

3. A fire fighter being arrested or other events that require immediate action by the fire department are classified as _____ incidents.

4. An effective manager should know what is going on and address most issues _____ they become major problems.

5. The sooner the fire officer receives bad news or becomes aware of problems, the sooner he or she can implement _____ action.

6. Employees who do not feel their input is valuable will _____ passing vital information to the fire officer.

7. One of the constraints to _____ is that the process assumes the problem statement is accurate and the criteria are valid.

8. A factor in deciding on the best solution to a problem is the core _____ system of your department.

9. An implementation plan must include a(n) _____ to ensure that the goals are met.

10. If the original solution cannot be implemented, a different schedule, solution, or Plan _____ should be considered.

Short Answer

Complete this section with short written answers using the space provided.

1. Identify an in-house issue and develop an implementation plan to solve the problem.

   _____

   _____

   _____

   _____

2. Describe a time when a Plan B solution was needed in order to resolve an issue.

   _____

   _____

   _____

   _____

3. Describe an incident that was created due to interpersonal conflict within a crew. How could the incident have been prevented?

   _____

   _____

   _____

   _____

4. List the steps of your fire department's complaint investigation procedure.

_____

_____

_____

_____

5. Identify an emotionally driven complaint in your department. Describe how the incident could have been handled.

_____

_____

_____

_____

Fire Alarms

The following case studies will give you an opportunity to explore the concerns associated with becoming a fire officer. Read each case study and then answer each question in detail.

1. As a member of the apparatus committee for your fire department, the fire chief asks for a recommendation of where to place four new pieces of apparatus that will be arriving in seven months. The fire department recently passed a bond levy and will be building two new fire stations. Due to the condition of the current apparatus, the decision was already made to purchase new apparatus prior to the new facilities being completed and placing some apparatus into storage. After approximately an hour of discussion, it is clear that there is no consensus as to where to place the new apparatus and where to store the current apparatus. The discussions become heated and people are irritated with each other. What should be done to resolve the issue?

_____

_____

_____

_____

2. One evening you are sitting in the fire station watch office and are approached by one of your fire fighters. He asks to speak with you in private about an issue that he needs your help with. The fire fighter explains to you that over the past two weeks, a couple of other members of the shift have been giving him a hard time while he is working out. He is larger than the other fire fighters on the shift and the comments have gotten bad enough that he no longer wants to work out while on shift. The department requires all members to work out a minimum of one hour per shift. After discussing the issue further, the fire fighter believes that it is almost harassment. He asks for your help in the situation. What do you do?

_____

_____

_____

_____

In-Basket

The following exercises will give you an opportunity to refine your writing skills.

1. Document a citizen complaint to initiate an internal investigation of an incident.

   _____

   _____

   _____

   _____

   _____

   _____

2. Outline a departmental problem or issue, complete with evidence, and provide a written proposal for policy development, or modification, to solve the issue.

   _____

   _____

   _____

   _____

   _____

   _____

3. Document an in-house incident involving a fist fight on the apparatus floor. Include all necessary information and a suggestion to remedy the situation.

   _____

   _____

   _____

   _____

   _____

   _____

Pre-Incident Planning and Code Enforcement

Workbook Activities

The following activities have been designed to help you. Your instructor may require you to complete some or all of these activities as a regular part of your fire officer training program. You are encouraged to complete any activity that your instructor does not assign as a way to enhance your learning in the classroom.

Chapter Review

The following exercises provide an opportunity to refresh your knowledge of this chapter.

Matching

Match each of the terms in the left column to the appropriate definition in the right column.

_____ 1. Catastrophic theory of reform

_____ 2. Construction type

_____ 3. Model codes

_____ 4. Ongoing compliance inspection

_____ 5. Standpipe system

_____ 6. Ordinance

_____ 7. Occupancy type

_____ 8. Automatic sprinkler system

_____ 9. Authority having jurisdiction

_____ 10. Regulation

A. The combination of materials used in the construction of a building or structure, based on the varying degrees of fire resistance and combustibility.

B. A law of an authorized subdivision of a state.

C. The purpose for which a building or a portion thereof is used or intended to be used.

D. A system of pipes with water under pressure that allows water to be discharged immediately when a sprinkler head operates.

E. Not a law, but it is written by a government agency and has the force of a law.

F. An organization responsible for enforcing the requirements of a code or standard.

G. Inspection of an existing occupancy to observe the housekeeping and confirm that the built-in fire protection features are in working order.

H. Codes generally developed through the consensus process with the use of technical committees developed by a code-making organization.

I. An arrangement of equipment located in a manner that allows water to be discharged through an attached hose and nozzles for the purpose of extinguishing a fire.

J. When fire prevention codes or firefighting procedures are changed in reaction to a fire disaster.

Multiple Choice

Read each item carefully and then select the best response.

_____ 1. Public education activities are often performed by:
   A. professionally trained educators.
   B. public information officers.
   C. fire officers.
   D. a combination of staff personnel and fire companies.

_____ **2.** A document developed by gathering data used by responding personnel to determine the resources and actions necessary to mitigate anticipated emergencies at a specific facility is a(n):
   A. hazardous use permit.
   B. pre-incident plan.
   C. ongoing compliance inspection.
   D. transcription.

_____ **3.** A six-step method for developing a pre-incident plan is provided in NFPA:
   A. 291.
   B. 1001.
   C. 1031.
   D. 1620.

_____ **4.** The pre-incident plan should evaluate the construction and physical elements of the structure and the site should occur in the:
   A. first step.
   B. second step.
   C. third step.
   D. fourth step.

_____ **5.** Determining which areas within the structure are resistant to the fire should occur in the:
   A. first step.
   B. second step.
   C. fourth step.
   D. sixth step.

_____ **6.** When planning for possible evacuations, the location and number of occupants should be established, along with a(n):
   A. media release.
   B. evacuation team.
   C. tracking system.
   D. transportation plan.

_____ **7.** The size of the building and its contents, construction type, occupancy, exposures, and fire protection system will affect the amount or number of:
   A. incident commanders.
   B. fire fighters in the RICO.
   C. water needed to extinguish the fire.
   D. departments required to handle the incident.

_____ **8.** In case responding personnel encounter hazardous materials, the pre-incident plan should provide contact information for the:
   A. hazardous materials coordinator.
   B. fire chief.
   C. media.
   D. workplace health and safety officer.

_____ **9.** In case responding personnel encounter high-voltage or specialized mechanical systems, the pre-incident plan should identify personnel who can provide:
    A. traffic control.
    B. resources and support.
    C. special equipment.
    D. technical assistance.

_____ **10.** The pre-incident plan should identify the appropriate incident management system to implement in the:
    A. second step.
    B. fourth step.
    C. fifth step.
    D. sixth step.

_____ **11.** The pre-incident plan should identify any special hazards in the:
    A. first step.
    B. second step.
    C. fourth step.
    D. sixth step.

_____ **12.** When a fire involves other objects and flashes over, it is in the:
    A. ignition phase.
    B. growth phase.
    C. fully developed phase.
    D. decay phase.

_____ **13.** A legally enforceable regulation that relates specifically to fire safety, such as the regulation of hazardous materials, is known as a(n):
    A. building code.
    B. fire prevention code.
    C. ordinance.
    D. law.

_____ **14.** When a jurisdiction passes an ordinance that adopts a specific edition of the model code, it is a(n):
    A. adoption by reference.
    B. regulation.
    C. adoption by transcription.
    D. fire prevention code.

_____ **15.** The model code is updated every:
    A. 6 months.
    B. year.
    C. 2 to 4 years.
    D. 5 to 6 years.

_____ **16.** In locations where freezing during cold weather occurs, automatic sprinkler systems are most often:
    A. deluge systems.
    B. wet-pipe systems.
    C. pre-action systems.
    D. dry-pipe systems.

_____ **17.** The standpipe system that provides both 1.5 and 2.5 connections is the Class:
    A. I.
    B. II.
    C. III.
    D. IV.

_____ **18.** The preferred extinguishing system from the 1960s to 1990s for computer and electronic rooms was the:
    A. carbon dioxide system.
    B. halon system.
    C. foam system.
    D. dry chemical system.

_____ **19.** The type of building construction that has structural elements made from either noncombustible or limited combustible materials is a:
   A. Type I.
   B. Type II.
   C. Type III.
   D. Type IV.

_____ **20.** The type of building construction that has noncombustible exterior walls and interior structural elements that are unprotected wood beams and columns with large cross-sectional dimensions is a:
   A. Type I.
   B. Type II.
   C. Type III.
   D. Type IV.

_____ **21.** The occupancy type of a building used for gatherings of people for entertainment, eating, or awaiting transportation is a(n):
   A. assembly.
   B. business.
   C. health care.
   D. residential.

_____ **22.** An occupancy used primarily for the storage or sheltering of goods or animals is a(n):
   A. mercantile.
   B. mixed.
   C. storage.
   D. unusual.

Fill-in

Read each item carefully, and then complete the statement by filling in the missing word(s).

1. A fire officer should always function as a(n) _____ with the community in a continuing effort to create a fire-safe community.

2. The investment that is made in the _____ phase results in a more valuable and effective response during an incident.

3. Conditions that would hamper access to an incident should be identified in the _____ step of pre-incident planning.

4. The _____ should identify the locations of any occupants who need assistance to evacuate a building safely.

5. In larger structures or high-occupancy structures, the _____ is often designed to shut down and work to eject smoke from the structure.

6. It is important to identify and understand the _____ system that feeds the hydrants in the area of the structure.

7. The total quantity of all combustible products that are within a room or space is referred to as the _____.

8. Similar to surveyor and architect drawings, the _____ plan shows the property from above, whereas the _____ plan shows the interior of the structure.

9. When a fire has consumed all of the oxygen but has retained the heat and has fuel, it has entered the _____ phase.

10. The construction of a new building, extension, or major renovation is regulated by the _____ code.

Short Answer

Complete this section with short written answers using the space provided.

1. Provide a list of the documentation included in your department's pre-incident plan.

_____

_____

_____

_____

2. Identify and provide contact information for the technical specialists and response teams within your department, as well as for the local technical specialists.

_____

_____

_____

_____

3. What is the scope of code enforcement authority delegated to fire officers in your jurisdiction?

_____

_____

_____

_____

4. Identify the five different building types within your community.

_____

_____

_____

_____

5. Identify buildings that represent the different occupancy types.

_____

_____

_____

_____

6. Describe the procedure used in your department for residences and buildings that do not meet code enforcement issues within your jurisdiction.

_____

_____

_____

_____

Fire Alarms

The following case studies will give you an opportunity to explore the concerns associated with becoming a fire officer. Read each case study and then answer each question in detail.

1. Your company has been assigned the task of updating current pre-incident plans as well as developing new pre-incident plans for all new commercial construction. The local college has just completed construction of a new classroom facility. They have invited you to come and conduct a walk-through. You decide that the walk-through would be an excellent time to create a pre-incident plan for the new facility. How should you proceed?

   _____

   _____

   _____

   _____

   _____

   _____

2. You arrive at the local elementary school to conduct a fire inspection. You contacted the school administrator last week and advised him of the inspection and scheduled an acceptable time to conduct the inspection. Upon arrival, you divide your company and start to inspect the large school. This is the largest facility that your company has inspected so far and it takes much longer than you anticipated. During your inspection, you find several minor code violations and document them appropriately. At the rear of the school, you find that two exit doors are chained shut while school is in session. What do you do?

   _____

   _____

   _____

   _____

   _____

   _____

In-Basket

The following exercises will give you an opportunity to refine your writing skills.

1. Develop pre-incident plans for your personal dwelling.

   _____

   _____

   _____

   _____

   _____

   _____

2. Prepare an internal memorandum to prepare fire fighters for their duties, assignments, and expectations for neighborhood residence inspections.

   _____

   _____

   _____

   _____

   _____

   _____

3. Complete an inspection/correction report for a local business.

_____

_____

_____

_____

_____

_____

4. An out-of-town industry has taken up occupancy in an old factory building and has changed the building's interior to meet its needs. Describe how you would pre-plan an old building with a new occupancy.

_____

_____

_____

_____

_____

_____

Budgeting and Organizational Change

Workbook Activities

The following activities have been designed to help you. Your instructor may require you to complete some or all of these activities as a regular part of your fire officer training program. You are encouraged to complete any activity that your instructor does not assign as a way to enhance your learning in the classroom.

Chapter Review

The following exercises provide an opportunity to refresh your knowledge of this chapter.

Matching

Match each of the terms in the left column to the appropriate definition in the right column.

_____ 1. Revenues

_____ 2. Supplemental budget

_____ 3. Bond

_____ 4. Expenditures

_____ 5. Fire tax district

_____ 6. Line-item budget

_____ 7. Base budget

_____ 8. Fiscal year

_____ 9. Budget

_____ 10. Cost recovery

A. A certificate of debt issued by a government or corporation that guarantees payment of the original investment plus interest by a specified future date.

B. Budget format where expenditures are identified in a categorized line-by-line format.

C. The level of funding that would be required to maintain all services at the currently authorized levels.

D. Proposed increases in spending for providing additional services.

E. When a fire department invoices a carrier or agency for extraordinary or special costs associated with a response to an incident.

F. An itemized summary of estimated or intended expenditures for a given period.

G. Twelve-month period for which an organization plans to use its funds.

H. The income of a government from all sources appropriated for the payment of the public expenses.

I. Special service district created to finance the fire protection of a designated district.

J. The act of spending money for goods or services.

Multiple Choice

Read each item carefully and then select the best response.

_____ 1. Fire department budgets identify the funds that are allocated to the fire department to define the:
   A. equipment purchases.
   B. services that the department will be able to provide.
   C. hiring and staffing mandate.
   D. importance of the fire department.

_____ **2.** Often new programs or budget items must identify where the funds would be found by:
    A. decreasing services or programs.
    B. decreasing revenues or increasing expenditures.
    C. increasing revenues or decreasing expenditures.
    D. maintaining services and programs.

_____ **3.** The United States Census summarizes the sources of state and local government tax revenues in:
    A. four general areas.
    B. five general areas.
    C. seven general areas.
    D. nine general areas.

_____ **4.** The Assistance to Firefighters Grant Program distributes:
    A. money from the state sales tax.
    B. federal funds to local jurisdictions.
    C. municipal taxes to fire departments.
    D. resources to family members of fire department personnel.

_____ **5.** The first step in developing a grant program proposal is to:
    A. identify the program need.
    B. determine if the grant program covers your problem area.
    C. approach municipal administrators.
    D. discuss the proposal with fire department members.

_____ **6.** A grant application has to describe how the fire department's needs fit the:
    A. priorities of the grant program.
    B. timelines of the grant program.
    C. administrative process of the funding body.
    D. fire department's mission statement.

_____ **7.** The format of a fire department's budget should comply with the recommendations of the:
    A. local elected officials.
    B. state (provincial) guidelines.
    C. Governmental Accounting Standards Board.
    D. legislated department auditors.

_____ **8.** In career departments, personnel expenses usually account for more than:
    A. 90% of the annual budget.
    B. 75% of the annual budget.
    C. 60% of the annual budget.
    D. 50% of the annual budget.

_____ **9.** The day-to-day delivery of service expenditures is covered in the:
    A. personnel budget.
    B. capital expenditures budget.
    C. municipal budget.
    D. operating budget.

_____ **10.** Mandated continuing training, such as hazardous materials and cardiopulmonary resuscitation recertification classes, are expenses within the:
   A. training budget.
   B. operating budget.
   C. capital budget.
   D. personnel budget.

_____ **11.** Construction, renovations, or expansion of municipal buildings are expensive projects that have funds allocated to them in the:
   A. training budget.
   B. operating budget.
   C. capital budget.
   D. personnel budget.

_____ **12.** Replacing employees with contracted employees is also referred to as:
   A. decentralizing.
   B. privatization.
   C. freelancing.
   D. personalization.

_____ **13.** Reducing duplication in staff and services by combining departments is:
   A. centralizing.
   B. focused staffing.
   C. hybridization.
   D. consolidation.

_____ **14.** A method of ensuring that there are sufficient funds available in a budget account to cover an estimated amount of a purchase is the use of a:
   A. purchase order.
   B. tender.
   C. bond.
   D. fire officer account.

_____ **15.** When the exact amount of a large purchase is not known, the item is purchased through a stringent process of a(n):
   A. special item expenditure.
   B. requisition.
   C. operating bond.
   D. bidding proposal.

_____ **16.** A fire department uses a _____ to give a vendor general information on what is desired, allowing the vendor to determine how it will meet the need.
   A. request for proposal.
   B. mobile data computer.
   C. purchase order.
   D. requisition.

Fill-in

Read each item carefully, and then complete the statement by filling in the missing word(s).

1. Budget preparation is both a(n) _____ and a(n) _____ process.

2. The _____ budget is the level of funding that addresses inflation adjustments, salary increases, and other predictable costs, so that the current level of operation would be maintained.

3. One of the primary roles of elected officials is to allocate government resources in a way that best _____ the public.

4. The mix of revenues available to local governments varies considerably from state to state because _____ governments set the rules for local governments.

5. Many volunteer fire departments are organized as independent _____ corporations.

6. Gathering information about a program, resource, or initiative is referred to as a(n) _____ assessment.

7. The _____ system allows budget analysts and financial managers to keep track of expenditures throughout a government budget.

8. Pension fund contributions, worker's compensation, and life insurance are included in the _____ category.

9. Durable items that last more than one budget year and are usually expensive are items in the _____ expenditures.

10. Annual budget costs fall into two areas: _____ and _____.

Short Answer

Complete this section with short written answers using the space provided.

1. Describe the budget cycle and key players involved in the development of your fire department's annual budget.

   _____

   _____

   _____

   _____

2. Identify the relationship with your fire department in regard to local and state government funding allocations.

   _____

   _____

   _____

   _____

3. Identify, describe, and provide last year's totals for the areas used in your fire department's budget to track expenditures.

   _____

   _____

   _____

   _____

4. Include copies of or describe the process used to access petty cash, purchase orders, requisitions, or tender purchases in your department.

   _____

   _____

   _____

   _____

5. Identify the areas of revenue included in your annual budget.

_____

_____

_____

_____

6. Identify the areas that your fire department would reduce if there were slight-to-major budget reductions due to declining revenues.

_____

_____

_____

_____

Fire Alarms

The following case studies will give you an opportunity to explore the concerns associated with becoming a fire officer. Read each case study and then answer each question in detail.

1. You have submitted numerous budget requests only to have them turned down due to the lack of fiscal resources. You and your fellow officers have identified many pieces of equipment that need to be replaced and no longer meet current national standards. After a recent staff meeting, you volunteer to look at grant opportunities to provide the necessary fiscal resources to purchase some of the identified needs of the department. The grant that immediately comes to mind is the FIRE Act grant program offered through the Department of Homeland Security. What is the four-step method to develop a competitive grant proposal?

_____

_____

_____

_____

_____

_____

2. As part of your department's strategic plan, it has been identified that there is a significant need for a new headquarters facility and two new pumpers in order to maintain and improve service delivery to the community in the future. The department does not have the fiscal resources to pay for such improvements. You have been asked by the fire chief to investigate the possibility of going to the citizens with a bond levy. What is a bond levy?

_____

_____

_____

_____

_____

_____

In-Basket

The following exercises will give you an opportunity to refine your writing skills.

1. Following your municipal and department regulations, develop a proposal and outline the key components to a fire department fundraiser that could be held in your community.

   _____

   _____

   _____

   _____

   _____

   _____

2. Following your state's guidelines, develop a grant proposal for a specific training program or department equipment need.

   _____

   _____

   _____

   _____

   _____

   _____

3. Prepare a grant proposal for replacement of your department's SCBA.

   _____

   _____

   _____

   _____

   _____

   _____

4. Prepare an annual budget for your fire department training academy. Include all personnel costs and operating expenses.

   _____

   _____

   _____

   _____

   _____

   _____

Fire Officer Communications

Workbook Activities

The following activities have been designed to help you. Your instructor may require you to complete some or all of these activities as a regular part of your fire officer training program. You are encouraged to complete any activity that your instructor does not assign as a way to enhance your learning in the classroom.

Chapter Review

The following exercises provide an opportunity to refresh your knowledge of this chapter.

Matching

Match each of the terms in the left column to the appropriate definition in the right column.

_____ 1. Environmental noise

    A. Short-term documents signed by the fire chief and lasting for a period of days to 1 year.

_____ 2. Company journal

    B. The path a message takes from and back to its original source.

_____ 3. Interrogatories

    C. A decision document prepared by the fire officer for the senior staff to support a decision or action.

_____ 4. Active listening

    D. A log book that records emergency, routine, or special activities and any liability-creating event, including special visitors to the fire station.

_____ 5. Communication cycle

    E. A written organizational directive that prescribes specific operational methods to be followed routinely for the performance of designated operations or actions.

_____ 6. Recommendation report

    F. A physical or sociological condition that interferes with the message in the communication process.

_____ 7. General orders

    G. A letter or report presented on fire department letterhead and signed by the chief officer or headquarters staff person.

_____ 8. Supervisor's report

    H. The ability to accurately interpret comments, concerns, and questions.

_____ 9. Standard operating procedures

    I. A form completed by an immediate supervisor after an injury or property damage accident.

_____ 10. Formal communication

    J. A series of formal written questions sent to the opposing side.

Multiple Choice

Read each item carefully and then select the best response.

_____ 1. Successful communication occurs when:

    A. a message is transferred quickly and efficiently.

    B. the desired outcome is achieved.

    C. two people exchange information and develop a mutual understanding.

    D. the message is internalized.

CHAPTER

14

_____ 2. Effective communication does not occur unless the intended message has been:
   A. sent and received.
   B. received and understood.
   C. understood with feedback.
   D. fully communicated.

_____ 3. The text of the communication—the information—is the:
   A. message.
   B. environment.
   C. feedback.
   D. medium.

_____ 4. The person in the communication cycle responsible for ensuring the message is formulated properly is the:
   A. receiver.
   B. interpreter.
   C. medium.
   D. sender.

_____ 5. When sending a message, any behavior is up for interpretation, so we must be careful to send only the intended message and attempt to:
   A. control the receiver.
   B. focus the sender.
   C. minimize misinterpretations.
   D. increase the number of mediums.

_____ 6. The receiver is the person who receives and:
   A. controls the medium.
   B. interprets the message.
   C. hears the message.
   D. sets the learning environment.

_____ 7. The communication cycle is completed by the verification of the receiver's interpretation of the message, which is also referred to as the:
   A. medium.
   B. noise.
   C. evaluation.
   D. feedback.

_____ 8. For a supervisor to function effectively and accurately interpret concerns, he or she must be able to:
   A. listen.
   B. schedule time to meet with subordinates.
   C. develop good messages.
   D. mediate discussions.

_____ 9. The purpose of active listening is to help the fire officer to:
   A. interact with the fire fighter.
   B. provide appropriate feedback.
   C. understand the fire fighter's viewpoint.
   D. control the communication.

_____ **10.** A good method to keep a conversation on topic is to:
   A. use hand gestures.
   B. make eye contact.
   C. pause between speakers.
   D. use direct questions.

_____ **11.** The informal communication system evident in all organizations, which is often based on incomplete facts and half-truths, is the:
   A. vacuum.
   B. grapevine.
   C. coffee row.
   D. smoking area.

_____ **12.** Emergency communications require:
   A. noiseless environments.
   B. various mediums for each message.
   C. the direct approach.
   D. primary and secondary senders.

_____ **13.** A letter on official fire department letterhead intended for someone outside the fire department is usually a(n):
   A. formal communication.
   B. informal communication.
   C. general order.
   D. legal correspondence.

_____ **14.** Formal documents that address a specific subject, policy, or situation, which are signed by the fire chief and are often used to announce promotions or personnel transfers, are:
   A. recommendation reports.
   B. supervisor's reports.
   C. announcements.
   D. general orders.

_____ **15.** In any situation where the fire department's reports or documentation are requested:
   A. the documents must be released as public information.
   B. the fire department's legal counsel should be consulted.
   C. a fire officer should be present during the review of the documents.
   D. the request should be refused respectfully and politely.

_____ **16.** A permanent reference kept by the fire officers that documents daily activities is referred to as a log book or a(n):
   A. company journal.
   B. officer's log.
   C. shift report.
   D. supervisor's report.

_____ **17.** Every type of emergency response must have some type of:
   A. officer's report.
   B. recommendation report.
   C. incident report.
   D. resource report.

_____ **18.** The keyboard, monitor, and printer are computer:
   A. software.
   B. hardware.
   C. internal components.
   D. local networks.

Fill-in

Read each item carefully, and then complete the statement by filling in the missing word(s).

1. Communication is a(n) _____ process that occurs in repetitive cycles.

2. The tone of voice or the look that accompanies a message can _____ the receiver's interpretation of the meaning.

3. The medium refers to the _____ that is used to convey the information from the sender to the receiver.

4. If _____ of the components of the communication process is missing, communication does not occur.

5. Anything that interferes with a message completing the communication cycle is _____.

6. When using a radio at an emergency scene, the messages at a minor incident should sound the _____ as the messages at a major incident.

7. A(n) _____ is intended to provide information to someone who needs or at least desires the information.

8. One purpose of the _____ report is to identify any personnel or resource shortages as soon as possible after the on-duty personnel report for duty.

9. Fire officers must clearly understand the purpose in order to write a report that is _____ and presents the necessary information in an understandable format.

10. A fire department can reach a large audience, at virtually no cost, by using a well-prepared _____.

Short Answer

Complete this section with short written answers using the space provided.

1. Identify three of your strengths and three of the challenges you face in trying to be an effective communicator.

_____

_____

_____

_____

2. Describe a situation where your prejudice or bias created a sociological environmental noise situation.

_____

_____

_____

_____

3. How do you handle members of your fire department who have been identified as the recurring sources of negative grapevine conversations?

_____

_____

_____

_____

4. Identify the main sources of noise that affect your meeting and training facilities.

_____

_____

_____

_____

5. List and describe your fire department's standard operating procedures for radio communications at emergency incidents.

_____

_____

_____

_____

Fire Alarms

The following case studies will give you an opportunity to explore the concerns associated with becoming a fire officer. Read each case study and then answer each question in detail.

1. After arriving at the scene of a structure fire at an apartment complex and taking command, you advise the first incoming truck to set up for a trench cut operation on Side B. You begin working on other duties and assigning other crews in accordance with your incident action plan. Your interior crew calls and advises that conditions are not improving and visibility is zero. You look up to find out what the ventilation operation is doing and you notice that the company is actually on Side D. What is your course of action?

_____

_____

_____

_____

_____

_____

2. You are the incident commander at the scene of a dormitory fire at the local college. You have several crews working on fire attack when there is a report from Division C of some unusual noise coming from above them. Division C believes it sounds like an imminent collapse and recommends abandoning the structure. What do you do?

_____

_____

_____

_____

_____

In-Basket

The following exercises will give you an opportunity to refine your writing skills.

1. Referencing an incident report, prepare written notes for an oral presentation.

_____

_____

_____

_____

_____

_____

2. Prepare a news release for the last hazardous materials or business structural fire your department mitigated.

_____

_____

_____

_____

_____

_____

3. Prepare a news release for a new technical rescue team that your fire department has placed in service.

_____

_____

_____

_____

_____

_____

4. Your battalion chief has requested a written explanation for an injury of a fire fighter on your pumper. The injury was sustained while the fire fighter was extinguishing a car fire. The fire fighter was not wearing gloves (as required) and cut his hand during overhaul. Prepare your report, explaining the injury and the actions you will take to prevent this type of injury in the future.

_____

_____

_____

_____

_____

_____

Managing Incidents

Workbook Activities

The following activities have been designed to help you. Your instructor may require you to complete some or all of these activities as a regular part of your fire officer training program. You are encouraged to complete any activity that your instructor does not assign as a way to enhance your learning in the classroom.

Chapter Review

The following exercises provide an opportunity to refresh your knowledge of this chapter.

Matching

Match each of the terms in the left column to the appropriate definition in the right column.

_____ 1. Incident commander	**A.** A supervisory level established to divide an incident into geographical areas of operations.
_____ 2. Branch	**B.** A supervisory level established to divide the incident into functional areas of operation.
_____ 3. Group supervisor	**C.** Command level in which objectives must be achieved to meet the strategic goals.
_____ 4. Task force	**D.** A specific function in which resources are assembled in an area at or near the incident scene to await instructions or assignments.
_____ 5. Staging	**E.** Any combination of single resources assembled for a particular tactical need, with common communications and a leader.
_____ 6. Strike team	**F.** Command level that entails the overall direction and goals of an incident.
_____ 7. Group	**G.** A supervisory position in charge of a functional operation at the tactical level.
_____ 8. Task level	**H.** Either a geographical or a functional assignment.
_____ 9. Tactical level	**I.** Specified combinations of the same kind and type of resources, with common communications and a leader.
_____ 10. Divisions	**J.** Command level in which specific tasks are assigned to companies.
_____ 11. Strategic level	**K.** A supervisory level established in either the operations or logistics function to provide a span of control.
_____ 12. Unit	**L.** The person who is responsible for all decisions relating to the management of the incident and who is in charge of the incident site.

Multiple Choice

Read each item carefully and then select the best response.

_____ 1. The organization developed to create a more effective system to deal with major incidents, through the major fire agencies in southern California, is the:
   A. FGC.
   B. NIMS.
   C. FIRESCOPE.
   D. FRP.

_____ 2. The Standard on an Emergency Services Incident Management System is NFPA:
   A. 1003.
   B. 1021.
   C. 1561.
   D. 1600.

_____ 3. If an incident escalates to the point that it requires a federal response, all of the participating agencies are expected to use the:
   A. ICS model.
   B. FGC model.
   C. FEMA model.
   D. NIMS model.

_____ 4. The responsibility and maintenance of the National Response Framework (NRF) belongs to:
   A. DHS.
   B. OSHA.
   C. NIOSH.
   D. IFSAC.

_____ 5. The federal agency responsible for conducting research and making recommendations for the prevention of work-related injury and illness is:
   A. FEMA.
   B. OSHA.
   C. NIOSH.
   D. IFSAC.

_____ 6. The "two-in-two-out" rule:
   A. only applies for IDLH atmosphere incidents.
   B. evolved into the rapid intervention team concept.
   C. is only used during structure fires.
   D. should be monitored by EMS personnel.

_____ 7. The command structure for an incident should only be:
   A. made up of fire department personnel.
   B. made up of fire officers.
   C. put in place when directed by the fire chief.
   D. as large as the incident requires.

8. The three strategic priorities of life safety, incident stabilization, and property conservation are the responsibility of the:
A. command staff.
B. safety officer.
C. fire chief.
D. incident commander.

9. The level of the command structure where the physical work is actually accomplished is the:
A. strategic level.
B. tactical level.
C. task level.
D. staff level.

10. The initial incident commander remains in charge of the incident until the situation is stabilized and terminated or until:
A. command is transferred.
B. a full IMS is in place.
C. the local authority having jurisdiction arrives.
D. the incident escalates.

11. Activating the command process includes announcing that command has been established and:
A. designating a safety officer.
B. initiating response orders.
C. providing an initial radio report.
D. receiving support from the command post.

12. When an emergency incident is so large that immediate establishment of command by the first-arriving officer is required, the decision has been to go to a(n):
A. investigation mode.
B. command mode.
C. fast-attack mode.
D. strategic mode.

13. The only member of the command staff who has the authority to stop or suspend operations, besides the incident commander, is the:
A. safety officer.
B. liaison officer.
C. operations section chief.
D. logistics section chief.

14. The responsibility of gathering and releasing incident information to the news media and appropriate agencies belongs to the:
A. administration officer.
B. planning section chief.
C. public education officer.
D. public information officer.

Fill-in

Read each item carefully, and then complete the statement by filling in the missing word(s).

1. Every fire officer should be prepared to function in a variety of roles in the _____ _____ system.

2. Legal sanctions accelerated adoption of fire fighter _____ practices.

3. The major difference in the two systems is that the FGC was designed for implementation in _____ situations, whereas IMS was designed for _____ -scale operations.

4. The _____ is a core set of doctrines, concepts, principles, terminology, and organizational processes.

5. _____ ruled in 1996 that fire fighters working within a structure fire were operating in an IDLH atmosphere.

6. For emergency operations, a recommended span of control is _____ to _____ individuals reporting to one supervisor.

7. Command of an incident should only be transferred to improve the _____ of the command organization.

8. The _____ and the _____ of an incident will determine how much of the management structure needs to be put into place.

9. Strategic objectives of how the emergency operations will be conducted are outlined and stated in the _____.

10. When the incident commander needs information obtained, managed, and analyzed, he or she will activate the _____ section.

Short Answer

Complete this section with short written answers using the space provided.

1. Outline and describe your responsibilities as the first-arriving officer at an emergency scene.

_____

_____

_____

_____

2. Describe the standard operating procedures for your department's tactical level of the incident management system.

_____

_____

_____

_____

3. Describe the incident management system used in your local area. Highlight any special guidelines or standard operating procedures.

_____

_____

_____

_____

4. Provide or outline the incident command worksheet used in your department.

_____

_____

_____

_____

5. Discuss a positive and a negative learning experience from different postincident critiques you have attended.

_____

_____

_____

_____

Fire Alarms

The following case studies will give you an opportunity to explore the concerns associated with becoming a fire officer. Read each case study and then answer each question in detail.

1. Fire units are currently working on the scene of a fire involving three units of a six-unit apartment building. You have just arrived. Your first priority is to meet up with the captain who established incident command to get a situation report. After the face-to-face with the captain, you conduct your own size-up and check on the status of crews and the stage of the fire. After conducting the size-up, you talk with the captain and assume command. You retain the captain at the command post as your aide. One problem that you find is that the captain says he did not have time to set up a personnel accountability system yet, so you do not really have any idea how many companies and personnel are on scene. What should be your next action?

_____

_____

_____

_____

_____

_____

2. Upon dispatch, Chief North pulls out of the station and notices the large thermal column to the west. The dispatch is for a fire in a strip mall on the west side of town. Based on the size of the column, Chief North requests a second alarm. Behind him are Engine 91 and Engine 94, with crews of the four each respectively. Upon arrival, they find a small laundromat in the strip mall with heavy smoke blowing out of Side A. Chief North knows that if they do not hit the fire hard and fast, it will spread quickly to the adjacent structures. What action could Chief North take to attempt to control the situation quickly?

_____

_____

_____

_____

_____

_____

In-Basket

The following exercises will give you an opportunity to refine your writing skills.

1. Prepare a written document for your last structure fire that could be referenced during a postincident review.

_____
_____
_____
_____
_____
_____

2. Develop speaker's notes or a lesson plan to present information on your incident management system to new fire department members.

_____
_____
_____
_____
_____
_____

3. Prepare an organizational chart showing your department's incident command structure for a residential fire.

_____
_____
_____
_____
_____
_____

4. Expand your created chart to include rescue, EMS, and HAZMAT involvement.

_____
_____
_____
_____
_____
_____

Fire Attack

Workbook Activities

The following activities have been designed to help you. Your instructor may require you to complete some or all of these activities as a regular part of your fire officer training program. You are encouraged to complete any activity that your instructor does not assign as a way to enhance your learning in the classroom.

Chapter Review

The following exercises provide an opportunity to refresh your knowledge of this chapter.

Matching

Match each of the terms in the left column to the appropriate definition in the right column.

_____ 1. Defensive attack   **A.** The location at which the primary logistics functions are coordinated and administered.

_____ 2. Personnel accountability report   **B.** A major division of the logistics section that oversees the communications, medical, and food units.

_____ 3. Incident action plan   **C.** Oversees the use of the elevators, operates the local building communication system, and assists in the control of the heating, ventilating, and air conditioning systems.

_____ 4. Service branch   **D.** A major division within the logistics section that oversees the supply, facilities, and ground support units.

_____ 5. Transitional attack   **E.** An advance into the fire building by fire fighters with hoselines or other extinguishing agents to overpower the fire.

_____ 6. Lobby control officer   **F.** Systematic method of accounting for all personnel at an emergency incident.

_____ 7. Offensive attack   **G.** The objectives reflecting the overall incident strategy, tactics, risk management, and member safety that are developed by the incident commander.

_____ 8. Support branch   **H.** When suppression operations are conducted outside the fire structure.

_____ 9. Base   **I.** A situation in which an operation is changing or preparing to change.

_____ 10. Stairwell support   **J.** This group moves equipment and water supply hoselines up and down the stairwells.

Multiple Choice

Read each item carefully and then select the best response.

_____ 1. The fire officer's personal physical involvement on an incident must:
   A. lead the company in all duties.
   B. always be the most efficient.
   C. never compromise his or her supervisory responsibilities.
   D. only occur if the subordinates are unable to complete the task.

_____ **2.** In addition to leading and participating in company-level operations, the fire officer should also be:
   A. evaluating the company's effectiveness.
   B. documenting the company's actions.
   C. directing EMS care as required.
   D. establishing command positions.

_____ **3.** During an emergency incident, the level of supervision provided by the fire officer should be balanced with the experience of the company members and the:
   A. size of the incident.
   B. nature of the assignment.
   C. number of departments responding.
   D. experience of the fire officer.

_____ **4.** The preferred style of leadership during emergency scene activities is:
   A. hierarchical.
   B. participative.
   C. paramilitary.
   D. authoritarian.

_____ **5.** Unlike nonemergency activities, emergency operations must be conducted:
   A. efficiently.
   B. in a very structured and consistent manner.
   C. by individual fire fighters.
   D. as quickly and timely as possible.

_____ **6.** A framework to allow activities at an emergency scene to be completed in an efficient manner and with all components of the organization working in concert with each other is provided by the:
   A. general operating guidelines.
   B. department regulations.
   C. standard operating procedures.
   D. command system.

_____ **7.** A good size-up that considers all the pertinent information, defines strategies and tactics, and assigns resources to complete those tactics is a(n):
   A. incident action plan.
   B. pre-incident plan.
   C. radio dispatch communication.
   D. incident commander order.

_____ **8.** One of the most significant factors in size-up is:
   A. location.
   B. time of response.
   C. company experience.
   D. visualization.

_____ **9.** According to Layman's Five-Step Size-Up Process, "things that are likely to happen or can be anticipated based on the known facts" are within the:
   A. facts step.
   B. probabilities step.
   C. situation step.
   D. decision step.

_____ **10.** According to Layman's Five-Step Size-Up Process, "the assessment of all three dimensions of the incident must be considered and factored into the incident action plan" occurs within the:
   A. facts step.
   B. probabilities step.
   C. situation step.
   D. decision step.

_____ **11.** The first phase of the National Fire Academy (NFA) Size-Up Process includes the:
   A. need to continually reassess the size-up as the situation evolves.
   B. question "What do I have?"
   C. pre-incident information.
   D. size and construction of the building.

_____ **12.** The mnemonic used by the National Fire Academy to assist in remembering the critical points that should be considered for a size-up is:
   A. WALLACE WAS HOT.
   B. RECEO VS.
   C. FOIL.
   D. ROGER OUT.

_____ **13.** Only when there are realistic benefits, like saving a life or valuable property, is it worth the risk of a(n):
   A. defensive attack.
   B. transition attack.
   C. rapid intervention attack.
   D. offensive attack.

_____ **14.** Unlike other fire flow formulas, the National Fire Academy fire flow formula is based on:
   A. cubic footage.
   B. gallons per minute.
   C. square footage.
   D. structure area.

_____ **15.** The highest priority in all incident action plans is:
   A. life safety.
   B. rescue and recovery.
   C. incident stabilization.
   D. property conservation.

_____ **16.** Minimizing water damage, using salvage covers, or venting smoke and heat from a building are characteristics of:
   A. life safety.
   B. rescue and recovery.
   C. incident stabilization.
   D. property conservation.

_____ **17.** The stage of RECEO VS that is designed to remove heat, smoke, and the products of combustion from a fire area is:
   A. extinguishments.
   B. ventilation.
   C. salvage.
   D. exposures.

_____ **18.** The fifth and last of the tactical priorities in the RECEO VS model is:
   A. ventilation.
   B. salvage.
   C. overhaul.
   D. confinement.

_____ **19.** Advancing hand lines to the room of origin, into stairways, into the attic, or to the floor above the fire is a task assignment usually associated with:
   A. extinguishing.
   B. exposure.
   C. ventilating.
   D. confinement.

_____ **20.** One consideration when an evacuation area is being established is the:
   A. time of day.
   B. number of resources available.
   C. nature of the event.
   D. type of neighborhood.

Fill-in

Read each item carefully, and then complete the statement by filling in the missing word(s).

1. The primary responsibility of a fire officer is to _____ the company while assigned tasks are being performed.

2. The fire officer is always responsible for relaying relevant _____ to his or her supervisor.

3. When performing their duties at an emergency incident, fire fighters often get tunnel vision; therefore, the fire officer must look at the big _____.

4. An inexperienced crew performing a high-risk task requires more _____ than a highly experienced crew performing a routine task does.

5. The specific size-up for an incident begins with the _____.

6. The final step in Layman's decision process is the _____, where the actual plan to mitigate the incident is developed.

7. The third phase of the National Fire Academy Size-Up Process includes the need for a(n) _____ analysis of the situation.

8. The key factor in selecting the appropriate strategic mode for an incident is the _____ analysis.

9. The volume of _____ must be sufficient to absorb the heat that is being released.

10. _____ are general and more equivalent to goals, whereas _____ are more specific, measurable, and like objectives that are used to meet the goals.

Short Answer

Complete this section with short written answers using the space provided.

1. Identify the size-up factors you would consider as the first-arriving officer to a structure fire.

_____
_____
_____
_____

2. Outline your department's standard operating procedures in establishing initial scene size-up.

_____
_____
_____
_____

3. What extra considerations must be made when supervising multiple companies?

_____

_____

_____

_____

4. Identify characteristics or factors that would indicate it is time to move from an offensive attack to a defensive attack.

_____

_____

_____

_____

5. Describe the scene safety considerations and standard operating procedures used at motor vehicle accidents.

_____

_____

_____

_____

Fire Alarms

The following case studies will give you an opportunity to explore the concerns associated with becoming a fire officer. Read each case study and then answer each question in detail.

1. Your engine company is dispatched for a commercial structure fire at Evergreen State College "C" Dorm. You know that based upon pre-incident plans, "C" Dorm does not have sprinklers and is a wood-framed, 4-story structure with about 80 residents. While en route, dispatchers advise that they are receiving several calls of a fire on the 3rd floor that is growing quickly. Dispatchers also advise that there are reports of several students trapped on the 4th floor. Your initial alarm assignment consists of three engine companies, one aid unit, and one truck. What do you do?

_____

_____

_____

_____

_____

2. You arrive as the acting battalion chief at the scene of a working fire in a 10-story commercial building. The fire is located on the 6th floor and the building has been evacuated. You already have four companies working their way to the fire floor and an additional four companies still en route to the call. You make contact with one of your engine company officers who is currently the incident commander. After an exchange of necessary information about the incident and what crews are doing so far, you conduct a scene size-up, and then assume command. What do you do next?

_____

_____

_____

_____

_____

_____

In-Basket

The following exercises will give you an opportunity to refine your writing skills.

1. Develop a standard operating procedure for a rapid intervention crew to follow in preparation for standby orders.

_____

_____

_____

_____

_____

_____

2. Recalling one emergency incident response in your career, develop an incident size-up using both the Layman's Five-Step and National Fire Academy Size-Up Processes.

_____

_____

_____

_____

_____

_____

3. Prepare a lesson plan detailing the virtues of the ability to read smoke in a structure fire.

_____

_____

_____

_____

_____

_____

Fire Cause Determination

Workbook Activities

The following activities have been designed to help you. Your instructor may require you to complete some or all of these activities as a regular part of your fire officer training program. You are encouraged to complete any activity that your instructor does not assign as a way to enhance your learning in the classroom.

Chapter Review

The following exercises provide an opportunity to refresh your knowledge of this chapter.

Matching

Match each of the terms in the left column to the appropriate definition in the right column.

_____ **1.** Trailer

A. Physical marks left on an object by the fire.

_____ **2.** Fire pattern

B. The exact physical location where a heat source and a fuel come in contact with each other and a fire begins.

_____ **3.** Char

C. Crime of maliciously and intentionally, or recklessly, starting a fire or causing an explosion.

_____ **4.** Accelerant

D. What the material is made of.

_____ **5.** Arson

E. Materials used to spread the fire from one area of a structure to another.

_____ **6.** Form of materials

F. The process of determining the origin, cause, development, and responsibility, as well as the failure analysis, of a fire or explosion.

_____ **7.** Point of origin

G. An item on which fire patterns are present, in which case the preservation is not for the item itself but for the fire pattern that appears on the item.

_____ **8.** Type of materials

H. An agent, often an ignitable liquid, used to initiate a fire or increase the rate of growth or spread of fire.

_____ **9.** Fire analysis

I. Carbonaceous material that has been buried and has a blackened appearance.

_____ **10.** Artifact

J. What the material is being used for.

Multiple Choice

Read each item carefully and then select the best response.

_____ **1.** The largest fire cause category is:
   A. incendiary.
   B. open flame.
   C. cooking.
   D. unknown.

_____ **2.** A qualified fire investigator has specialized training and, in most cases, is certified in accordance with NFPA:
   A. 1003.
   B. 1021.
   C. 1033.
   D. 1500.

_____ **3.** If a fire does not meet the criteria for calling a fire investigator, local policy usually dictates the investigation and report is completed by the:
   A. fire officer on scene.
   B. incident commander.
   C. fire chief.
   D. deputy chief of operations.

_____ **4.** Conduction, convection, and radiation are the three methods of:
   A. fire growth.
   B. ignition.
   C. char development.
   D. heat transfer.

_____ **5.** As a fire grows and develops, it follows the same pattern as the:
   A. fuel load.
   B. smoke and heat.
   C. air currents.
   D. oxygen and fuel sources.

_____ **6.** The origin of the fire is typically at the base of the V- or U-shaped pattern that is also known as the:
   A. origin pattern.
   B. growth pattern.
   C. movement pattern.
   D. development pattern.

_____ **7.** The specific cause of a fire can be determined only after:
   A. salvage and overhaul are complete.
   B. all potential causes have been identified.
   C. the char has been analyzed.
   D. evidence has been documented.

_____ **8.** The energy source that caused the material to ignite is the:
   A. source of ignition.
   B. fuel.
   C. accelerant.
   D. heat.

_____ **9.** Generation, transmission, and heating are the components of:
   A. incendiary fires.
   B. reliable fire fuels.
   C. a component ignition source.
   D. movement patterns.

_____ **10.** According to the U.S. Fire Administration, the leading cause of vehicle fires is:
   A. arson.
   B. revenge.
   C. vandalism.
   D. mechanical failure.

_____ **11.** Wildfires tend to spread vertically through:
   A. convection.
   B. conduction.
   C. radiation.
   D. wind.

_____ **12.** Any fire that is deliberately ignited under circumstances in which the person knows that the fire should not be ignited is a(n):
   A. accidental fire.
   B. deliberate fire.
   C. incendiary fire.
   D. undetermined fire.

_____ **13.** The chemical decomposition that results in a gradual lowering of the ignition temperature of the wood until autoignition occurs is:
   A. oxygenation.
   B. pyrolysis.
   C. decaying.
   D. fulgurites.

_____ **14.** Arson to an abortion clinic, religious institution, or ecologically damaging business is arson due to:
   A. extremism motives.
   B. vandalism motives.
   C. spite/revenge motives.
   D. crime concealment motives.

_____ **15.** A fire officer who conducts a preliminary cause investigation and suspects a crime has occurred should immediately:
   A. document the information he or she has found.
   B. secure the scene.
   C. remove emergency responders from the scene.
   D. request the response of a fire investigator.

_____ **16.** Tangible items that can be identified by witnesses are:
   A. documentary evidence.
   B. testimonial evidence.
   C. demonstrative evidence.
   D. artifacts.

_____ **17.** The process of recreating the physical scene before the fire occurred is:
   A. scene identification.
   B. fire scene reconstruction.
   C. point of origin isolation.
   D. ignition source revisiting.

_____ **18.** A company officer may be called on to testify as a _____.
   A. witness.
   B. victim.
   C. friend.
   D. complainant.

Fill-in

Read each item carefully, and then complete the statement by filling in the missing word(s).

1. Any fire that results in a serious injury or fatality meets the criteria for a(n) _____ investigation.

2. If the cause of the fire was _____, a crime has occurred and must be fully investigated.

3. The point of origin should be determined in the _____ step of the fire investigation.

4. The amount of heat that was transferred to the surrounding area and objects is indicated by the _____ pattern.

5. Determining the fuel that was first ignited should occur in the _____ step of cause determination.

6. The cause of a fire cannot be established until all _____ causes have been identified and considered and only one cannot be eliminated.

7. The use of _____ questions allow a witness to tell what he or she saw or knows.

8. When conducting a vehicle fire investigation, note the make, model, and year, as well as the _____.

9. A business owner or employee may elect to burn the business in order to destroy records that show _____.

10. A professional arsonist is likely to try to make a fire appear _____.

Short Answer

Complete this section with short written answers using the space provided.

1. List the criteria for initiating a formal fire investigation and for requesting the response of an investigator or investigator team in your jurisdiction.

_____
_____
_____
_____

2. List characteristics or indicators of incendiary fires.

_____
_____
_____
_____

3. Describe different methods used to secure a scene for further investigation.

_____
_____
_____
_____

4. Using your experiences, identify and describe an emergency incident that reflects the various motives for arson.

_____
_____
_____
_____

5. What tips and advice would you provide for a fellow officer prior to his or her first court appearance in a fire investigation?

_____
_____
_____
_____

Fire Alarms

The following case studies will give you an opportunity to explore the concerns associated with becoming a fire officer. Read each case study and then answer each question in detail.

1. You are dispatched to a residential structure fire in a 2-story wood-frame structure. Upon your arrival, you find fire in four separate locations in the house, with an exterior fire burning near the front door. During your scene size-up, you find graffiti sprayed across the rear of the house. Several small explosions are heard from within the structure. What do you do?

_____
_____
_____
_____
_____

2. Your fire company arrives at the scene of a reported fire at the local high school. You find signs of possible forced entry, but as you enter the school, there does not appear to be any problem. You break your crew into two teams to investigate. A short time later, you hear a small explosion behind you. You try to contact the other team on the radio with no answer. You quickly go with your team to find out what happened. You turn the next corner and see that flames are now shooting out of one of the classrooms and the other team is on the ground, dazed but uninjured. What happened?

_____
_____
_____
_____
_____

In-Basket

The following exercises will give you an opportunity to refine your writing skills.

1. After attending and taking part in a fire investigation, complete an investigation report.

_____
_____
_____
_____
_____
_____

2. Prepare a list of indicators and clues that a first-arriving officer and company should look for during the first five minutes of structural firefighting.

_____
_____
_____
_____
_____

Workbook Activities

The following activities have been designed to help you. Your instructor may require you to complete some or all of these activities as a regular part of your fire officer training program. You are encouraged to complete any activity that your instructor does not assign as a way to enhance your learning in the classroom.

Chapter Review

The following exercises provide an opportunity to refresh your knowledge of this chapter.

Matching

Match each of the terms in the left column to the appropriate definition in the right column.

_____ 1. Followership

A. The process of questioning a situation that causes concern.

_____ 2. Situational awareness

B. An individual's ability to perform tasks that require specific knowledge or skills.

_____ 3. Communication

C. Describes how commanders can recognize a plausible plan of action.

_____ 4. Inquiry

D. The process of evaluating the severity and consequences of an incident and communicating the results.

_____ 5. Personal competence

E. An individual's own internal strengths, capabilities, and character.

_____ 6. Crew resource management

F. Describes how commanders make decisions in their natural environment.

_____ 7. Technical competence

G. Leaders can be effective only to the extent that people are willing to accept their leadership.

_____ 8. Advocacy

H. The successful transfer and understanding of a thought from one person to another.

_____ 9. Naturalistic decision making

I. A behavior modification training system developed by the aviation industry to reduce its accident rate.

_____ 10. Recognition-primed decision making

J. The statement of opinion that recommends what one believes is the proper course of action under a specific set of circumstances.

Multiple Choice

Read each item carefully and then select the best response.

_____ 1. The National Institute for Occupational Safety and Health reports that a major contributing factor in a large proportion of fire fighter fatalities is:
   A. inefficient use of resources.
   B. lack of training.
   C. poor response times.
   D. human error.

_____ **2.** Gordon Dupont has identified that humans make mistakes based on one or a combination of:
  A. 5 reasons and ways.
  B. 8 reasons and ways.
  C. 12 reasons and ways.
  D. 15 reasons and ways.

_____ **3.** Avoidance, entrapment, and mitigating consequences are three methods for dealing with:
  A. errors.
  B. fire fighters.
  C. evidence.
  D. problem solving.

_____ **4.** For crew resource management (CRM) to become effective in a team:
  A. there must be training.
  B. all participants must document their progress.
  C. there must be a strong leader.
  D. every member must participate.

_____ **5.** When en route to an alarm, the entire crew should be exchanging only information that is pertinent to:
  A. the scene size-up.
  B. responding to and arriving safely at the scene of the alarm.
  C. their role and the tasks they will complete at the incident.
  D. the standard operating procedures.

_____ **6.** Crew resource management promotes subordinates being:
  A. intelligent.
  B. diligent to their assignments.
  C. proactive.
  D. efficient taskmasters.

_____ **7.** In Todd Bishop's five-step assertive statement process, "stating the concern" is the:
  A. first step.
  B. second step.
  C. third step.
  D. fourth step.

_____ **8.** In Todd Bishop's five-step assertive statement process, "stating a solution" is the:
  A. second step.
  B. third step.
  C. fourth step.
  D. fifth step.

_____ **9.** The three components of leadership are effective leadership, leadership skills, and:
  A. honesty.
  B. ability.
  C. integrity.
  D. trust and respect.

_____ **10.** The four critical areas for self assessment of one's ability to function as part of a team are mental condition, attitude, understanding human behavior, and:
   A. physical condition.
   B. respect.
   C. empathy.
   D. compassion.

_____ **11.** The dividing of responsibilities among individuals and teams in a manner that allows for effective accomplishment is:
   A. resource assignment.
   B. task allocation.
   C. delegation.
   D. activity responsibilities.

_____ **12.** A recommended practice for improving decision making includes gaining experience, training constantly, improving communication skills, and:
   A. knowing your resources.
   B. enhancing technical skills.
   C. preplanning incidents.
   D. reviewing past incidents.

Fill-in

Read each item carefully, and then complete the statement by filling in the missing word(s).

1. _____ are the unsafe acts committed by people who are in direct contact with the situation or system.

2. _____ conditions have two kinds of adverse effects: error-provoking conditions within the local workplace and long-lasting holes or weaknesses in the defenses.

3. _____ is an error-management model with three activities: avoidance, entrapment, mitigating consequences.

4. CRM suggests that developing a(n) _____ language and teaching appropriate _____ behavior are the keys to reducing errors resulting from miscommunication.

5. CRM advocates speaking directly and assertively, yet in a manner that does not challenge the _____ of a superior.

6. The factor to keep in mind is that communication should not focus on _____ is right, but on _____ is right.

7. The inquiry and advocacy process and the assertive statement are essential components of the _____ segment of CRM.

8. CRM promotes members working together for the _____.

9. An officer who fails to listen or refuses to listen is _____.

10. CRM emphasizes that in order for decision making to be efficient there has to be someone who has _____ for decisions and outcome.

Short Answer

Complete this section with short written answers using the space provided.

1. Identify strategies you use to maintain emergency scene situational awareness.

   _____

   _____

   _____

   _____

2. Describe the process used in your department's debriefing sessions.

   _____

   _____

   _____

   _____

3. Identify your strengths and challenges as a communicator while at an emergency incident.

   _____

   _____

   _____

   _____

4. How do you handle conflict between a subordinate and yourself?

   _____

   _____

   _____

   _____

5. What characteristics make you worthy of being "followed?"

   _____

   _____

   _____

   _____

Fire Alarms

The following case studies will give you an opportunity to explore the concerns associated with becoming a fire officer. Read each case study and then answer each question in detail.

1. Your fire company is dispatched to an ALS motor vehicle accident. Three members of your crew are new probationary fire fighters who have been out of the academy for three weeks. This is their first major vehicle accident and you can see the adrenaline in their eyes. The call is at the far edge of your jurisdiction and it will be a 15-minute response. On the way to the call, the fire fighters in the back are constantly talking on the headsets and you are having a difficult time hearing the radio. On top of that, your driver begins talking about his fishing trip from yesterday. What should you do?

_____

_____

_____

_____

_____

_____

2. During a practical training exercise, you take your engine company out to conduct some hose evolutions. While conducting one of the FDC evolutions, there is a disagreement over how the department expects the evolution to be completed. One of your more senior fire fighters disagrees with how you are having them conduct the evolution and says that you are not doing it per the department standard. You cut him off and are irritated by this challenge. You are sure that you are doing it correctly. You go back to the station and check in the hose evolution book and realize that you were doing it incorrectly and that the fire fighter was correct. What should you do?

_____

_____

_____

_____

_____

_____

In-Basket

The following exercises will give you an opportunity to refine your writing skills.

1. In a letter to your supervisor, identify the main human factors that have led to errors you have witnessed during emergency responses and provide suggestions on how to eliminate the errors.

_____

_____

_____

_____

_____

_____

2. Develop a mission statement model, identifying the main components of your philosophy on leadership.

_____

_____

_____

_____

_____

_____

3. Prepare a series of training drills designed to demonstrate the need and use of crew resource management, using real-life experiences and incidents.

_____

_____

_____

_____

_____

_____

Answer Key

Chapter 1: Introduction to the Fire Officer

Matching

1. G (page 10)
2. E (page 15)
3. H (page 10)
4. B (page 11)
5. A (page 12)
6. D (page 9)
7. C (page 20)
8. F (page 19)
9. J (page 11)
10. I (page 13)

Multiple Choice

1. C (page 4)
2. B (page 4)
3. D (page 4)
4. B (page 7)
5. A (page 7)
6. B (page 7)
7. C (page 8)
8. A (page 9)
9. D (page 10)
10. A (page 9)
11. A (page 10)
12. B (page 11)
13. C (page 11)
14. D (page 13)
15. C (page 14)
16. D (page 17)

Fill-in

1. incident management system (page 5)
2. career; volunteer (page 7)
3. fire officer (page 6)
4. Benjamin Franklin (page 7)
5. 2.5 (page 8)
6. George Smith (page 8)
7. Governments (page 10)
8. value (page 17)

Fire Alarms

1. The organized structure of a fire agency consists of the chain of command. The chain of command provides for a formalized structure so that the department can manage both day-to-day operations as well as emergency operations. The chain of command normally goes from fire fighter to lieutenant, from lieutenant to captain, from captain to battalion chief, from battalion chief to assistant or division chief, to the chief of the department. This chain of command should be followed by all members within the department when bringing forth information or requests.
2. The baby boomer generation is quickly reaching retirement age, which is resulting in a larger than normal amount of retirements from the fire service. Some departments are experiencing up to half of their workforce being able to retire soon. This will mean that more new fire officers will be needed in the coming years as well as more personnel with less experience.

Chapter 2: Preparing for Promotion

Matching

1. I (page 34)
2. D (page 27)
3. A (page 25)
4. C (page 27)
5. J (page 30)
6. H (page 31)
7. B (page 31)
8. F (page 27)
9. E (page 33)
10. G (page 33)

Multiple Choice

1. B (page 25)
2. C (page 25)
3. A (page 25)
4. D (page 26)
5. B (page 26)
6. C (page 26)
7. D (page 27)
8. C (page 30)
9. B (page 30)
10. A (page 30)
11. A (page 31)
12. B (page 26)
13. C (page 30)
14. A (page 30)
15. D (page 32)
16. B (page 30)

Fill-in

1. spoils (page 25)
2. needs (page 26)
3. changing (page 26)
4. class specification (page 27)
5. multiple choice (page 27)
6. in-basket (page 30)

Fire Alarms

1. First, it is important to closely read the testing announcement and to be very clear on what assessment center elements will be used. Wasting time on the elements that will not be tested will take away valuable time and effort. Next, ensure that you know what each element is and what it normally consists of. This includes conducting some background research on each of the elements that will be tested. In this instance, the assessment center will include an emergency incident simulation, in-basket exercise, and interpersonal interaction exercise. Remember, when studying for the assessment center, keep in mind what level of fire officer position you are applying for. This affects what may be expected in each element of the assessment center, i.e., a lieutenant assessment center may test you on your abilities as an engine company officer arriving at the scene of a fire while a battalion chief assessment center may test your abilities to command multiple companies at the scene of a fire. Another valuable tool includes mock exercises to run through to gain confidence and a level of comfort with each of the elements.

2. There are primarily two reasons that would cause a department to test for a fire officer position. The retirement of one or more officers in a department would create openings in the officer ranks that would need to be filled. These positions are often very important to the functionality and efficiency of a department. As a fire department grows and expands its role in the community by adding additional units and/or stations, there would be a need for additional fire officers that would also influence a department's need to test.

Chapter 3: Fire Fighters and the Fire Officer

Matching

1. F (page 50)
2. H (page 52)
3. I (page 55)
4. G (page 53)
5. B (page 48)
6. C (page 53)
7. D (page 54)
8. A (page 52)
9. J (page 50)
10. E (page 53)

Multiple Choice

1. C (page 42)
2. D (page 44)
3. A (page 45)
4. B (page 45)
5. D (page 45)
6. A (page 47)
7. C (page 46)
8. B (page 46)
9. A (page 49)
10. D (page 50)
11. C (page 50)
12. D (page 52)
13. A (page 53)
14. C (page 54)
15. C (page 44)
16. A (page 53)

Fill-in

1. beginning of shift (page 43)
2. emergency incidents (page 44)
3. decisions (page 44)
4. color (page 45)
5. responsibility (page 45)
6. fire officer (page 45)
7. less (page 46)
8. private (page 46)
9. presence (page 47)
10. authority (page 50)

Fire Alarms

1. Fire fighters technically have three options to report sexual harassment: they can start federally, at the local government level, or within the fire department. In this instance, a formal complaint has not technically been filed and the fire fighter has approached you as the fire officer to deal with the situation. The fire officer must function as the first stop in a multi-stop procedure. First and foremost, the fire officer must be familiar with the departments SOPs as they pertain to

sexual harassment. The fire officer may be required to conduct an initial investigation. Some of the guidelines to follow include: keep an open mind, treat the complainer with respect and compassion, do not blame the complainer, do not retaliate against the complainer, follow procedure, interview those involved, look for corroboration or contradiction, stay confidential, document everything, and cooperate with government agencies if necessary.

2. As the fire officer, it is your responsibility to enforce the rules of the department and those expected of you by the administration. While you disagree with the memo and the manner in which it was posted, you still have a responsibility to respect the assistant chief and enforce the rule. It would be necessary to remind the fire fighters of the memo and that they would be violating the assistant chief's directive by going to the store. You could further advise them that if they disagreed with the memo, they should take appropriate steps to work and get the directive changed. Until a change, the crew would need to abide by the directive. If you as the fire officer had an issue with the directive, it would be best to follow the appropriate action of bringing the issue up at a staff meeting or in private with the assistant chief.

Chapter 4: Understanding People: Management Concepts

Matching

1. B (page 64)
2. H (page 67)
3. J (page 68)
4. I (page 70)
5. A (page 64)
6. D (page 70)
7. E (page 70)
8. G (page 70)
9. F (page 65)
10. C (page 64)

Multiple Choice

1. A (page 63)
2. B (page 63)
3. A (page 66)
4. C (page 67)
5. C (page 67)
6. D (page 67)
7. C (page 68)
8. A (page 68)
9. D (page 68)
10. B (page 68)
11. A (page 71)
12. D (page 72)
13. A (page 62)
14. C (page 63)
15. B (page 71)
16. C (page 65)

Fill-in

1. Management (page 62)
2. directly (page 70)
3. mission statement (page 71)
4. efficiency (page 71)
5. delegating (page 72)
6. autonomy (page 64)
7. 5 (page 67)
8. skill (page 70)
9. benefits (page 70)
10. physiological (page 65)

Fire Alarms

1. A key to the problems that you are encountering may be that it is a very hot, dry day. In Maslow's hierarchy of needs, the physiological needs of your fire fighters are very important and are considered level one needs. Your crew's performance may be due to the heat and their lack of rest and rehydration. Give them a rehabilitation break and allow them to get some rest and rehydrate, and then try the evolutions again. You may see a profound change in their performance.

2. As a fire officer, it is always helpful to assist those new officers around you and to provide them a guiding hand from time to time. It is vital that the new fire officer work to build trust with his subordinates as quickly as possible. The officer's ability or inability to gain trust will make or break his career. To be a success, the new fire officer must know the job, be consistent, make his fire fighters feel strong, support them, and walk the talk.

Chapter 5: Organized Labor and the Fire Officer

Matching

1. F (page 90)
2. D (page 81)
3. I (page 79)
4. C (page 88)
5. A (page 82)
6. J (page 90)
7. E (page 87)
8. B (page 93)
9. G (page 79)
10. H (page 83)

Multiple Choice

1. B (page 79)
2. D (page 79)
3. C (page 79)
4. A (page 79)
5. A (page 79)
6. C (page 80)
7. B (page 79)
8. C (page 80)
9. B (page 80)
10. D (page 80)
11. C (page 82)
12. A (page 88)
13. D (page 89)
14. B (page 91)
15. A (page 91)
16. C (page 91)

Fill-in

1. lagged (page 80)
2. prohibit (page 83)
3. win-win (page 87)
4. all (page 79)
5. closed (page 81)
6. IAFF (page 81)
7. labor; management (page 85)
8. counterproductive (page 84)
9. organization; workers (page 90)
10. advocate (page 91)

Fire Alarms

1. As the fire officer, you should be courteous to the concerns of your shift personnel and understand their concerns about the change in working conditions/hours. As part of the grievance procedure, it is fairly common that the fire officer is the first "official" of the department to receive a grievance regarding a labor practice. In this case, as the fire officer, you would receive the grievance and submit it up the chain of command. As a side note, because of the labor contract, if there was language that stipulated a change in working conditions that would require negotiation, this grievance would most likely be in favor of the employees and such negotiations would need to occur before the shift starting time could be changed to 0800.

2. In looking at the success and joint labor–management cooperation that the rule-by-objective method provides, the issue should be discussed in an open labor management setting. The RBO model would dictate that major issues such as this should include analysis, a decision, education, implementation, revision, and review. Instead of creating an adversarial issue, the RBO would provide for open, positive discussion on the issue to include possible solutions and resolution that both parties could be happy with. It is important to note that RBO requires the department and union to jointly accept and adopt a win-win compromise.

Chapter 6: Safety and Risk Management

Matching

1. D (page 105)
2. K (page 115)
3. J (page 114)
4. H (page 108)
5. I (page 110)
6. G (page 114)
7. L (page 111)
8. A (page 107)
9. E (page 108)
10. F (page 117)
11. C (page 108)
12. B (page 106)

Multiple Choice

1. D (page 100)
2. D (page 99)
3. A (page 101)
4. C (page 102)
5. C (page 104)
6. D (page 104)
7. B (page 104)
8. C (page 105)
9. A (page 105)
10. D (page 106)
11. B (page 106)
12. C (page 106)
13. A (page 106)
14. A (page 108)
15. D (page 108)
16. D (page 111)
17. B (page 111)
18. A (page 114)

Fill-in

1. Safe; safety (page 99)
2. Prevention (page 99)
3. 1000 (page 111)
4. reduce (page 102)

5. teams; accountability (page 103)
6. seat belts (page 103, 104)
7. changing; unanticipated (page 105)
8. multiple (page 106)

Fire Alarms

1. A proactive fire officer can improve the quality of the shift's health and safety. A healthy diet is an excellent first step to making a stronger, more competitive shift. Exercise is another important aspect to improving fire fighters' efficiency and longevity. Because driving accounts for such a large amount of a fire fighter's time, an equal amount of emphasis should be placed on the importance of driver safety and emergency vehicle operations. Wearing seat belts should be second nature. Finally, safety on the fire ground is a responsibility that ultimately leads back to the fire officer. Crews should be aware and alert at all times when it comes to dealing with tasks on emergency scenes. Accidents happen all too often when fire fighters become complacent. Fire officers should make a point of training their crews to react to their jobs on the fire ground as individual safety officers. Regular training during shifts and reminders of standard department safety practices are preventive proactive maintenance.

2. In most events, the fire department's health and safety officer will be charged with investigating the accident. However, the initial investigation may be delegated to the shift officer. The two goals of the accident investigation should be ensuring that all required documentation is complete and identifying any corrective actions needed to prevent another injury, accident, or incident from occurring.

Chapter 7: Training and Coaching

Matching

1. G (page 122)
2. F (page 123)
3. H (page 124)
4. I (page 124)
5. B (page 129)
6. C (page 124)
7. J (page 122)
8. A (page 122)
9. E (page 125)
10. D (page 124)

Multiple Choice

1. C (page 123)
2. B (page 124)
3. D (page 124)
4. C (page 125)
5. D (page 125)
6. D (page 125)
7. A (page 126)
8. C (page 128)
9. B (page 128)
10. A (page 128)
11. B (page 129)
12. C (page 129)
13. B (page 129)
14. A (pages 129; 132)
15. B (page 134)
16. D (page 125)

Fill-in

1. consensus (page 123)
2. qualifications (page 123)
3. qualifications; certificates (page 123)
4. deficiency (page 125)
5. simple; complex (page 126)

6. feedback (page 128)
7. realistic (page 128)
8. competency; existing (page 129)
9. trainee or new fire fighter (page 132)
10. live; flammable/combustible; one (page 133)

Fire Alarms

1. The fire officer should follow the four-step method when developing such a course. The four-step method includes preparation, presentation, application, and evaluation. During the preparation phase, the fire officer should obtain the necessary information and develop a lesson plan that meets the desired objectives of the course. The presentation phase includes the actual lecture and/or instructional phase of the training. Application provides the opportunity for the class participants to demonstrate their knowledge of the material learned and the final evaluation phase includes an assessment of the student's progress and knowledge of the material, either written or practical.

2. Accreditation is a process in which a peer group evaluates a department or program to a standard set of criteria. This peer group makes the determination if a department's program is worthy of accreditation. Two organizations currently provide accreditation to fire service professionals. They are the International Fire Service Accreditation Congress (IFSAC) and the National Board on Fire Service Professional Qualifications (Pro Board).

Chapter 8: Evaluation and Discipline

Matching

1. F (page 153)
2. I (page 148)
3. J (page 153)
4. D (page 153)
5. H (page 147)
6. G (page 146)
7. B (page 153)
8. A (page 151)
9. E (page 146)
10. C (page 153)

Multiple Choice

1. D (page 142)
2. C (page 142)
3. D (page 143)
4. A (page 143)
5. B (page 144)
6. D (page 145)
7. A (page 145)
8. B (page 145)
9. C (page 147)
10. B (page 147)
11. D (page 148)
12. A (page 148)
13. A (page 148)
14. B (page 148)
15. C (page 151)
16. B (page 151)
17. A (page 153)
18. D (page 154)

Fill-in

1. job performance, on-duty behavior, and problem resolution (page 142)
2. dismissal (page 142)
3. goals (page 142)
4. job description (page 143)
5. 4 (page 144)
6. documented (page 145)
7. T-account (page 145)
8. before (page 146)
9. period of time (page 146)
10. lenient (page 146)

Fire Alarms

1. Lieutenant Novak could maintain a performance log on each person under his command. This documentation would prove to be invaluable when preparing for annual performance evaluations. This performance log would allow him to maintain a record of the fire fighter's activities by date, along with a brief description of his or her performance.

2. The assistant chief could approach this in a couple different ways; however, the most successful would be to approach it using positive discipline. He could point out the discrepancies in the evolution and provide them with some feedback of what needs to improve. An option would be to retest the company immediately or provide them with some additional "corrective" time. By using positive discipline, the assistant chief would ensure that the company has a positive outcome.

Chapter 9: Leading the Fire Company

Matching

1. D (page 163)
2. E (page 165)
3. J (page 162)
4. G (page 164)
5. F (page 161)
6. H (page 164)
7. B (page 163)
8. I (page 164)
9. A (page 164)
10. C (page 162)

Multiple Choice

1. A (page 164)
2. D (page 165)
3. C (page 164)
4. D (page 163)
5. B (page 163)
6. A (page 163)
7. C (page 169)
8. A (page 169)
9. B (page 170)
10. C (page 165)
11. D (page 164)
12. A (page 165)
13. B (page 165)
14. D (page 164)
15. C (page 169)
16. C (page 170)

Fill-in

1. specific; attainable (page 165)
2. responsibility (page 165)
3. increase (page 164)
4. want (page 162)
5. situation (page 163)

6. followers (page 172)
7. expectancy (page 165)
8. communication (page 165)
9. external; internal (page 164)
10. lowest (page 163)

Fire Alarms

1. An excellent way to monitor a shift's mood, understand their work style, and obtain useful input is by using the National Fire Academy's 360° feedback system. By using this system, no single person is ever singled out and everyone gets equal treatment whether it be criticism or praise. This is a straightforward way of bringing people's strengths and weaknesses to the table without being brash and offensive. Hearing things firsthand or reading about other people's perception of you is an excellent way to learn and grow as a fire officer. When members of the crew see that you have taken their comments and suggestions to heart and made a conscious effort to utilize them, they will respect your authority even more.

2. The fire officer is a very busy individual in a first-arriving unit. By taking control of the situation in a calm, controlled manner, the fire officer creates a professional, organized mood for the emergency workers on scene. In addition to the individuals already on scene, arriving units and personnel can expect a well-organized and mitigated scene when they do arrive. This will prove to be constructive for communications and will smooth out any operational issues as well.

Chapter 10: Working in the Community

Matching

1. D (page 177)
2. H (page 187)
3. F (page 180)
4. J (page 188)
5. A (page 180)
6. I (page 186)
7. B (page 182)
8. C (page 181)
9. E (page 184)
10. G (page 182)

Multiple Choice

1. D (page 176)
2. C (page 177)
3. C (page 177)
4. A (page 177)
5. B (page 178)
6. A (page 180)
7. D (page 180)
8. B (page 181)
9. C (page 181)
10. D (page 182)
11. D (page 182)
12. A (page 184)
13. B (page 184)
14. B (page 189)
15. A (page 188)
16. A (page 188)

Fill-in

1. needs; characteristics (page 177)
2. preventing (page 178)
3. community-based (page 176)
4. sensitivity (page 177)
5. culture (pages 178–180)

6. empowers/encourages (page 180)
7. contact (page 181)
8. local (page 181)
9. spokesperson (page 188)
10. quoted (page 189)

Fire Alarms

1. Before beginning your project, you must address all of the issues that your fire chief wants you to cover to make this event a success. This includes the type of audience you will be addressing, the age of the people who will be visiting the fire department's booth, the volume of visitors, and the concerns that the public may have about any specific regulations or ordinances. After obtaining this information, it will be much easier for you to plan for the task.

   Information before an event such as this can make the difference between calling the event a success or a failure. For example, expecting one hundred visitors and really having two hundred visitors could deplete your staff, handouts, and flyers.

2. Even though this is an unfortunate situation to deal with, you can use this incident as a teaching tool to instill the importance of giving accurate information to the dispatcher when reporting an emergency. After the fire has been suppressed and overhauled, explain to the onlookers that your engine company was dispatched to a completely different location and that this lengthened the incident response time considerably. Also issue a press release to further spread the educational message to the public and remove any doubts about your department's competence and response abilities.

Chapter 11: Handling Problems, Conflicts, and Mistakes
Matching

1. E (page 203)   3. H (page 200)   5. D (page 201)   7. A (page 200)   9. F (page 203)
2. I (page 207)   4. G (page 196)   6. J (page 196)   8. C (page 196)   10. B (page 198)

Multiple Choice

1. C (page 196)   5. D (page 197)   9. D (page 198)   13. A (page 199)   17. D (page 204)
2. B (page 196)   6. A (page 197)   10. B (page 199)   14. A (page 200)   18. B (page 205)
3. A (page 197)   7. C (page 198)   11. C (page 200)   15. C (page 201)   19. C (page 207)
4. C (page 197)   8. B (page 198)   12. B (page 200)   16. A (page 203)   20. A (page 207)

Fill-in

1. problem (page 196)   6. stop (page 198)
2. conflicts or complaints (page 196)   7. brainstorming (page 199)
3. high-profile (page 197)   8. value (page 199)
4. before (page 197)   9. schedule (page 200)
5. corrective (page 198)   10. B (page 200)

Fire Alarms

1. The committee chair needs to call for a break and needs to bring the committee back together with some clearly defined rules to move the process forward. A process needs to be developed using the five-step decision-making process. Everyone needs to have the opportunity to provide input, but valid input is a must and the decisions need to be logical.

2. It is very important as a fire officer to completely listen to the complaint/concern, ask questions, and take notes so that you clearly understand the concern before making any assumptions. The next step is to investigate the complaint to obtain any necessary information either through observation or interviewing the two fire fighters who have been harassing the concerned fire fighter. After the investigation phase, you must take action to correct the problem. The action taken must follow department policy without prejudice and be professional. Follow-up with the complainant should occur to close the situation and make sure the issue has been resolved.

Chapter 12: Pre-Incident Planning and Code Enforcement
Matching

1. J (page 220)   3. H (page 221)   5. I (page 223)   7. C (page 228)   9. F (page 220)
2. A (page 225)   4. G (page 239)   6. B (page 221)   8. D (page 222)   10. E (page 225)

Multiple Choice

1. D (page 215)
2. B (page 215)
3. D (page 216)
4. A (page 216)
5. B (page 217)
6. C (page 218)
7. C (page 218)
8. A (page 218)
9. D (page 218)
10. C (page 219)
11. C (page 218)
12. C (page 219)
13. B (page 220)
14. A (page 222)
15. C (page 222)
16. D (page 223)
17. C (page 223)
18. B (page 224)
19. B (page 227)
20. D (page 227)
21. A (page 228)
22. C (page 230)

Fill-in

1. partner (page 231)
2. pre-incident (page 219)
3. first (page 216)
4. pre-incident plan (page 218)
5. HVAC (page 218)
6. distribution (page 218)
7. fuel load (page 219)
8. plot; floor (page 220)
9. decay (page 219)
10. building (page 220)

Fire Alarms

1. Prior to conducting the pre-incident plan survey, it is imperative that you contact the building owner, in this case the college facilities supervisor, and obtain permission to complete the pre-incident plan survey. NFPA 1620 should be used as a guide to completing the pre-incident plan survey. Prior to conducting the pre-incident plan survey, make sure to develop an action plan with your crew and assign members to collect specific information about the structure. This will make the process go smoothly when you arrive at the facility. Conduct the pre-incident plan survey in a professional manner and ensure that you are able to obtain all of the necessary information. Normally, it is most convenient to collect all of the data in one trip. However, due to the size of the facility, you may need to make two or more trips. After collecting data, place them in your department's pre-incident plan format and submit for approval. To make pre-incident plans a valuable resource, make sure that each shift reviews new pre-incident plans.

2. This is a major code violation and needs to be addressed *immediately* with the school. The school representative should be contacted as soon as possible, made aware of the violation, and corrective action must be taken immediately. This should be done in a professional manner while explaining the seriousness of the violation. The code violation should be referenced on the inspection form with proper documentation, including a digital photo, and should be forwarded to the fire marshal's office. This type of violation should be checked for future compliance on a regular basis due to the safety issue.

Chapter 13: Budgeting and Organizational Change

Matching

1. H (page 245)
2. D (page 246)
3. A (page 251)
4. J (page 245)
5. I (page 247)
6. B (page 249)
7. C (page 246)
8. G (page 245)
9. F (page 245)
10. E (page 249)

Multiple Choice

1. B (page 245)
2. C (page 247)
3. D (page 247)
4. B (page 248)
5. B (page 248)
6. A (page 248)
7. C (page 249)
8. A (page 249)
9. D (page 249)
10. A (page 250)
11. C (page 251)
12. B (page 252)
13. D (page 252)
14. A (page 255)
15. B (page 255)
16. A (page 255)

Fill-in

1. technical; political (page 245)
2. base (page 246)
3. serves (page 247)
4. state (page 247)
5. nonprofit (page 247)
6. needs (page 248)
7. accounting (page 249)
8. personnel (page 249)
9. capital (page 250)
10. continuity; personnel (page 251)

Fire Alarms

1. The four-step method to develop a competitive grant proposal includes: 1) conducting a community and fire department needs assessment; 2) comparing weaknesses to the priorities of the grant program; 3) deciding what to apply for; 4) completing the application.

2. A bond levy is a certificate of debt issued by a government or corporation guaranteeing payment of the amount plus interest by a specific future date. This funding has to be approved by the citizens of the jurisdiction through a referendum. Bonds are normally repaid over a period of ten to thirty years and provide a fixed rate of return to any investors. Bonds are similar to a mortgage in which the facility or equipment is paid for while it is being used. Bonds are repaid through the collection of a special property tax or future income.

Chapter 14: Fire Officer Communications

Matching

1. F (page 270)
2. D (page 273)
3. J (page 273)
4. H (page 269)
5. B (page 267)
6. C (page 278)
7. A (page 273)
8. I (page 277)
9. E (page 272)
10. G (page 272)

Multiple Choice

1. C (page 267)
2. B (page 267)
3. A (page 267)
4. D (page 268)
5. C (page 268)
6. B (page 268)
7. D (page 269)
8. A (page 269)
9. C (page 269)
10. D (page 269)
11. B (page 270)
12. C (page 271)
13. A (page 272)
14. D (page 273)
15. B (page 273)
16. A (page 273)
17. C (page 275)
18. B (page 279)

Fill-in

1. circular (page 267)
2. influence (page 268)
3. method (page 268)
4. one (page 269)
5. noise (page 270)
6. same (page 271)
7. report (page 273)
8. morning (page 273)
9. accurate (page 278)
10. news release (page 289)

Fire Alarms

1. There was evidently a breakdown in communications. The breakdown most likely occurred in the initial instructions due to the communications process not being followed. If the truck company officer had provided feedback to the incident commander to ensure that the information was received and understood, the mistake in location could have been averted.

2. Without hesitation, you should follow your department's SOPs for abandoning a structure and immediately remove crews from the hazardous area, check accountability, regroup, and evaluate the situation. With emergency communications, it is important that you be direct, speak clearly, use a normal tone of voice, speak in plain English, and try to avoid being in the area of other noise sources. In this instance, it is important to remain calm and maintain accountability of all personnel in accordance with your department's SOPs.

Chapter 15: Managing Incidents

Matching

1. L (page 290)
2. K (page 290)
3. G (page 293)
4. E (page 299)
5. D (page 298)
6. I (page 299)
7. B (page 297)
8. J (page 290)
9. C (page 290)
10. A (page 297)
11. F (page 290)
12. H (page 297)

Multiple Choice

1. C (page 287)
2. C (page 288)
3. D (page 288)
4. A (page 288)
5. C (page 293)
6. B (page 294)
7. D (pages 294–295)
8. D (page 290)
9. C (page 290)
10. A (page 293)
11. C (page 290)
12. B (page 291)
13. A (page 295)
14. D (page 295)

Fill-in

1. incident management (page 287)
2. accountability (page 293)
3. routine; large (page 287)
4. National Incident Management System (NIMS) (page 288)
5. OSHA (page 293)
6. 3; 5 (page 290)
7. quality (page 293)
8. size; complexity (page 293)
9. incident action plan (IAP) (page 295)
10. planning (page 295)

Fire Alarms

1. Fire-ground accountability is paramount to any successful incident and must be established from the beginning. In this case, that did not happen and as the incident commander, you need to correct that problem immediately to ensure that you can account for how many companies and personnel are at the incident. Depending upon the incident management system used in your department, sending the captain to collect accountability tags would be the first step. Most likely, you will have to go to each company to do this. If a company is in an interior combat role, it makes it difficult to gather tags. Gather their backup tags off their apparatus until they exit the structure. Once you have all company tags, you can account for the personnel that you have working on the scene.

2. In order to attempt a quick knockdown and control of the fire with the limited company availability, a fast attack or deck gun attack would be warranted. In order to control and knock down this fire, large enough water delivery must be provided in order to overcome the BTUs being produced by the fire. An effective deck gun attack with the first engine using its deck gun through the front of the structure and using three-quarters of its onboard tank water while establishing a water supply should make a significant impact on knocking down the fire. When the fire is knocked down to a manageable size, then appropriately sized hand lines can be moved in for extinguishment.

Chapter 16: Fire Attack

Matching

1. H (page 315)
2. F (page 324)
3. G (page 312)
4. B (page 326)
5. I (page 315)
6. C (page 326)
7. E (page 315)
8. D (page 326)
9. A (page 326)
10. J (page 326)

Multiple Choice

1. C (page 310)   **6.** C (page 311)   **11.** C (page 314)   **16.** D (page 326)
2. A (page 310)   **7.** A (page 312)   **12.** A (page 315)   **17.** B (page 327)
3. B (page 310)   **8.** D (page 312)   **13.** D (page 315)   **18.** C (page 327)
4. D (page 311)   **9.** B (page 313)   **14.** C (page 325)   **19.** D (page 328)
5. B (page 311)   **10.** C (page 313)   **15.** A (page 325)   **20.** C (page 329)

Fill-in

1. supervise (page 328)
2. information (page 310)
3. picture (page 310)
4. direct supervision (page 310)
5. dispatch (page 312)
6. plan of operation (page 313)
7. ongoing (page 314)
8. risk; benefit (page 314)
9. applied water (page 317)
10. Strategies; tactics (page 319)

Fire Alarms

1. As the first-arriving fire officer, it is imperative to keep a clear mind and maintain control. An incident of this magnitude with a serious life threat can become overwhelming quickly. Your first priorities should be rescue, confinement, and suppression. This type of fire would dictate an upgrade to an additional alarm prior to arriving on the scene based upon the information provided by dispatch. An aid in size-up would include the mnemonic "WALLACE WAS HOT."

2. As the incident commander, your major priorities should be personnel accountability, developing the IAP, and setting up a command structure. A high-rise fire scene needs additional command positions.

Chapter 17: Fire Cause Determination

Matching

1. E (page 341)   **3.** I (page 334)   **5.** C (page 342)   **7.** B (page 333)   **9.** F (page 336)
2. A (page 348)   **4.** H (page 341)   **6.** J (page 335)   **8.** D (page 335)   **10.** G (page 344)

Multiple Choice

1. C (page 333)   **5.** B (page 334)   **9.** C (page 335)   **13.** B (page 339)   **17.** B (page 344)
2. C (page 333)   **6.** C (page 334)   **10.** D (page 337)   **14.** A (page 343)   **18.** A (page 344)
3. A (page 333)   **7.** B (page 335)   **11.** A (page 338)   **15.** D (page 333)
4. D (page 334)   **8.** A (page 335)   **12.** C (page 338)   **16.** C (page 344)

Fill-in

1. formal (page 332)
2. deliberate (page 333)
3. first (page 333)
4. intensity (page 334)
5. second (page 335)
6. potential (page 335)
7. open-ended (page 336)
8. VIN (vehicle identification number) (page 338)
9. embezzlement (page 342)
10. accidental (page 342)

Fire Alarms

1. The first action would be to maintain safety of all personnel working on the scene and take a defensive posture. If possible, it would be advisable to take photos of the structure as quickly as possible to provide a record of the situation found, including the graffiti on the back of the house. Law enforcement and the fire marshal should be advised to respond and assist in any subsequent investigation. It is important not to jump to any conclusions. As the fire officer, it is important to protect as much evidence as possible with as little disruption as is necessary. Following the fire, it is important to have your personnel complete fire fighter observation reports, secure the scene, and deny entry. Depending on your department's procedures, you may or may not play a role in the investigation process and assist the fire investigators.

2. Most likely, the fire fighters were involved in a backdraft created in the classroom. A possible cause of the backdraft could have been from someone pouring excessive fuel on the floor. In this situation, the fire would flash but then die down due to insufficient oxygen. When air was introduced to the area, the fire could reignite. This is most likely what occurred to the fire fighters when they opened the door of the classroom.

Chapter 18: Crew Resource Management

Matching

1. G (page 357)
2. D (page 364)
3. H (page 354)
4. A (page 354)
5. E (page 356)
6. I (page 352)
7. B (page 356)
8. J (page 354)
9. F (page 358)
10. C (page 358)

Multiple Choice

1. D (page 364)
2. C (page 353)
3. A (page 353)
4. D (page 354)
5. B (page 354)
6. C (page 354)
7. B (page 356)
8. C (page 356)
9. D (page 356)
10. A (page 357)
11. B (page 357)
12. C (page 358)

Fill-in

1. Active failures (page 353)
2. Latent (page 353)
3. Crew resource management (CRM) (page 353)
4. standard; assertive (page 354)
5. authority (page 354)
6. who; what (page 354)
7. communication (page 356)
8. common good (page 356)
9. dangerous (page 357)
10. responsibility (page 357)

Fire Alarms

1. It is important that the entire crew focus on the call that they are responding to—not only to prepare, but to allow you to focus on the radio information and the driver to focus on the road. The crew should only exchange information that is pertinent to responding to and arriving safely at the scene.

2. It is important that you recognize that the fire fighter was correct and make the correction with the shift. It is not bad to be wrong, but it is bad if you do not recognize that you were wrong to the fire fighter and crew. Your credibility with your crew will be damaged if you do not admit the misinformation. Additionally, it is important when dealing with people that you do not cut them off when disagreeing with them, but respect their perspective and effectively listen to them.